***FM 3-04.111**

Field Manual
No. 3-04.111

Headquarters
Department of the Army
Washington, D.C., 7 December 2007

Aviation Brigades

Contents

Distribution Restriction: Approved for public release; distribution is unlimited.

***This publication supersedes FM 3-04-111, 21 August 2003.**

Figures

Tables

Preface

Doctrine provides a military organization with unity of effort and a common philosophy, language, and purpose. This document is the Army's keystone doctrine for fighting and sustaining aviation brigades.

This field manual (FM) is intended for all aviation commanders, staffs, and any United States (U.S.) military personnel expecting to conduct operations with Army aviation units. The operational concepts described in this manual reinforce fundamental principles found in Army doctrine. It provides overarching doctrinal guidance for employing aviation brigades in full spectrum operations and a foundation for developing tactics, techniques, and procedures (TTP) in other/follow-on Army manuals.

FM 3-04.111 expands employment doctrine for aviation units and describes considerations for forming aviation battalion task forces (ABTFs).

- Chapter 1 focuses on aviation brigade organizations and missions and provides aviation brigade fundamentals for reader consideration.
- Chapter 2 provides doctrine for organizing operations' command and control (C2) systems. It provides the organization for command post (CP) structure and general guides for roles and functions of CPs and their cells. Chapter 2 also describes duties of brigade leaders and staff.
- Chapter 3 discusses employment of the aviation brigade, planning considerations, and air-ground integration.
- Chapter 4 focuses on the formation and employment of ABTFs. It provides considerations for task organization based on mission and theater, and discussion of aviation task force employment principles.
- Chapter 5 discusses aviation brigade maintenance and logistics providing insight regarding the need to optimize mission accomplishment while balancing sustainability, mobility, and survivability. This chapter also focuses on the aviation support battalion's (ASB's) mission and function.

The appendices provide guidance, procedures and recommendations concerning specific subjects and amplify concepts referred to in the chapters. Appendix A discusses the Army Force Generation Model. Appendix E provides basic aircraft characteristics.

This FM applies to aviation forces through all operational themes—peacetime military engagements to major combat operations—and full spectrum operations—offensive, defensive, stability, and civil support.

This manual applies to the Active Army, the Army National Guard/Army National Guard of the United States, and the United States Army Reserve unless otherwise stated. Procedures described herein are intended as a guide and are not to be considered inflexible. Each situation in combat must be resolved by an intelligent interpretation and application of the doctrine set forth herein.

Finally, FM 3-04.111 furnishes a foundation for subordinate doctrine, force design, materiel acquisition, professional education, and individual and unit training.

The proponent of this publication is United States Army Training and Doctrine Command (TRADOC). Send comments and recommendations on Department of the Army (DA) Form 2028 (Recommended Changes to publications and Blank Forms) or automated link (http://www.usapa.army.mil/da2028/daform2028.asp) to Commander, United States Army Aviation Warfighting Center (USAAWC), ATTN: ATZQ-TD-D, Fort Rucker, Alabama 36362-5263. Comments may be e-mailed to the Directorate of Training and Doctrine (DOTD) at av.doctrine@us.army.mil. Other doctrinal information can be found on the Internet at Army Knowledge Online (AKO) https://us.army.mil/suite/page/389908

This publication has been reviewed for operations security (OPSEC) considerations.

Chapter 1

Missions and Organization

Aviation brigades are organized and equipped to support Army combined arms operations as well as joint, interagency, and multinational (JIM) operations. This chapter describes the mission and organization of each type of aviation brigade, and the fundamentals common to all aviation brigades. The description for each type of aviation brigade is based on the official table of organization and equipment (TOE). Operationally, all units are resourced according to the modified table of organization and equipment (MTOE). Actual organizations may vary from TOEs described in this field manual.

SECTION I – OVERVIEW

Contents

1-1. The aviation brigade is organized to synchronize operations of multiple aviation battalions simultaneously. Because the brigade is modular and tailorable, it can task organize as required for reconnaissance, security, close combat attack (CCA), interdiction attack, air assault, and air movement operations in support of ground forces.

1-2. Aviation brigades must rely on realistic training and established standing operating procedures (SOPs) to facilitate task organization and the addition or subtraction of subordinate units.

1-3. Throughout this manual, the term company includes troop and battalion includes squadron. Where appropriate the terms troop and squadron may be used when specifically discussing attack reconnaissance squadrons (ARSs), air cavalry squadrons (ACSs), or their respective subordinate units.

THE SPECTRUM OF CONFLICT

1-4. The spectrum of conflict, described in FM 3-0, ranges from stable peace to general war. See figure 1-1, page 1-2. Overlapping operational themes that occur along the spectrum are—

- Peacetime military engagement.
- Limited intervention.
- Peace operations.
- Irregular warfare.
- Major combat operations (MCO).

1-5. Aviation brigades execute various missions along this spectrum according to the operational theme characterized by the situation at the time. Generally, more than one operational theme is active at a given time, causing differences in the way the missions are executed, according to appropriate rules of engagement (ROE) and rules of interaction (ROI).

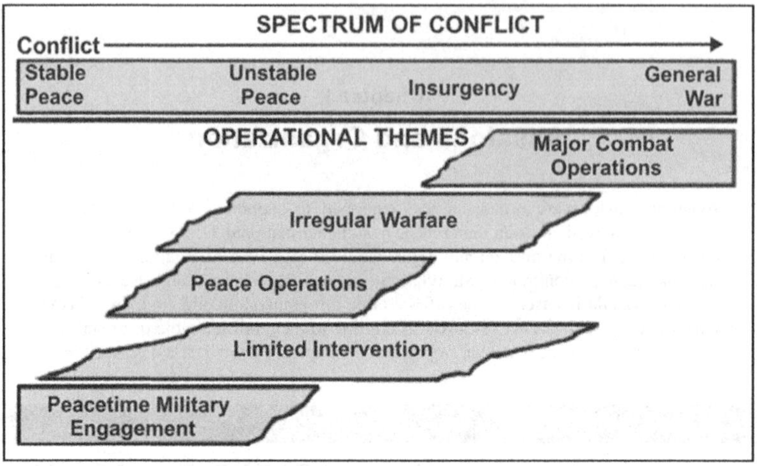

Figure 1-1. The spectrum of conflict and operational themes

1-6. Operational themes may change over time for various reasons including planned phases; changes caused by friendly, enemy, or neutral activity; revised political guidance; and unexpected opportunities. Operational themes should not be confused with tactical tasks or activities. As a rule, operational themes are too general for assigning missions. Rather, they describe the general characteristics of the major operation, not the details of execution.

Components of Full Spectrum Operations

1-7. There are four components of full spectrum operations—offensive, defensive, stability, and civil support.

1-8. **Offensive operations** defeat and destroy enemy forces, and seize terrain, resources, and population centers. They impose the commander's will on the enemy and achieve decisive victory.

1-9. **Defensive Operations** defeat an enemy attack, buy time, economize forces, or develop conditions favorable for offensive or stability operations

1-10. **Stability Operations** restore, establish, preserve, or exploit security and control over areas, populations, and resources. Stability operations are executed outside the U.S.

1-11. **Civil Support Operations** are conducted to address the consequences of natural or manmade disasters, accidents, and incidents within the U.S. and its territories.

Combat Power

1-12. Combat power is the actual application of force; the conversion of fighting potential into effective action. Combat power is derived from the six warfighting functions (WFFs) tied together by leadership. Aviation brigades combine WFFs to generate combat power. Table 1-1, page 1-3, provides examples of tasks aviation brigades conduct to support each WFF.

Table 1-1. Aviation brigade's role in Army warfighting functions

Army Warfighting Function	*Aviation Brigade's Role*
Movement & Maneuver	Support ground maneuver elements in contact through CCA. Conduct air assault in support of search and attack operations. Conduct movement to contact to locate and destroy enemy forces.
Intelligence	Conduct area reconnaissance to identify adequate routes and locate bypasses. Perform surveillance to confirm or deny enemy activity.
Fires	Utilize attack reconnaissance helicopters to conduct battle damage assessment (BDA) of fires. Designate for laser-guided artillery or other service munitions during joint air attack team (JAAT) operations.
Sustainment	Perform aircraft recovery to include insertion of downed aircraft recovery teams (DARTs) and ground maintenance contact teams. Support forward arming and refueling point (FARP) emplacement and resupply operations. Perform casualty evacuation (CASEVAC) and aeromedical evacuation (MEDEVAC).
C2	Provide battle command on the move (BCOTM). Provide retransmission capability to air and ground commander. Provide air traffic services (ATS).
Protection	Provide convoy security. Conduct area security through counter mortar and rocket operations.

CONCEPT OF OPERATIONS

1-13. The aviation brigade participates in all three operations that are expressed in the commander's concept (refer to FM 3-0). These operations include—

- **Decisive operations** that directly accomplish the mission assigned by higher headquarters and conclusively determine the outcome of major operations, battles or engagements.
- **Shaping operations that** create and preserve conditions for a successful decisive operation.
- **Sustaining operations** that enable the above two operations by generating and maintaining combat power.

SECTION II – MISSIONS

1-14. The role of the aviation brigade is to conduct and/or support ground maneuver through aviation operations. The brigade must prepare to fight as a whole, support brigade combat teams (BCTs) using pure or task-organized units, and conduct multiple independent missions requiring pure or task-organized units. Each aviation brigade is tailored for specific missions; however, each accepts other organizations and performs missions not necessarily defined in the TOE mission statement. Aviation brigade missions include—

- Reconnaissance.
- Security.
- Movement to contact.
- Attack.
- Air assault.
- Air movement.
- C2 support.

- Aeromedical evacuation.
- Casualty Evacuation (CASEVAC).
- Personnel Recovery (PR) operations.

1-15. The aviation brigade is also capable of conducting enabling missions to support operations and facilitate regeneration of combat power. These enabling missions include—

- Downed aircraft recovery.
- FARP operations.
- Aviation maintenance.
- Air Traffic Services (ATS).

1-16. All combat aviation brigades (CABs) have the capability to perform as a BCT when reinforced with appropriate ground units. These brigades can perform screen missions without augmentation, and can perform guard and cover missions when properly reinforced.

SECTION III – ORGANIZATION

1-17. There are six distinct types of aviation brigades. These brigades can work directly with supported maneuver units as a brigade or by forming aviation task forces (TFs) for specific missions for specific periods of time. The aviation brigades are—

- CAB (Heavy).
- CAB (Medium).
- CAB (Light).
- CAB (Expeditionary)
- Theater Aviation Brigade.
- Theater Aviation Brigade (Composite).

HEAVY, MEDIUM, AND LIGHT COMBAT AVIATION BRIGADES

1-18. These CABs are organized along the same construct varying only by the number and type of ARBs. The base construct for a CAB is two ARBs, an attack helicopter battalion (AHB), general support aviation battalion (GSAB), and ASB. Each brigade contains a headquarters and headquarters company (HHC) providing personnel and equipment for brigade C2 functions, and security and defense for the brigade CP.

1-19. Heavy and medium CABs have more robust firepower capability, while the light CABs require smaller-sized United States Air Force (USAF) lift to deploy ARS helicopters. The divisions that have medium CABs have a forced entry mission, requiring a mix of flexibility, speed, and firepower.

1-20. When fully fielded, each CAB will have one unmanned aircraft system (UAS) company.

1-21. Figure 1-2 (page 1-5) illustrates the organization of the heavy, medium and light CAB.

COMBAT AVIATION BRIGADES (EXPEDITIONARY)

1-22. The CAB (Expeditionary [E]) is designed primarily to focus on homeland security operations. These operations include counter-drug missions, humanitarian assistance, disaster relief, civil disturbance, counterterrorism, and domestic support, of which the National Guard is uniquely suited to perform. Elements of the CAB (E) can be task organized for deployment abroad.

1-23. The Army National Guard (ARNG) CAB (Expeditionary) (figure 1-3, page 1-6) is similar to the heavy aviation brigade except for a security and support battalion (S&S BN) (in lieu of one attack reconnaissance battalion [ARB]) currently equipped with OH-58Cs and, when fielded, the light utility helicopter (LUH). The ARB is resourced at 16 AH-64s on hand, while the GSAB has 8 CH-47s.

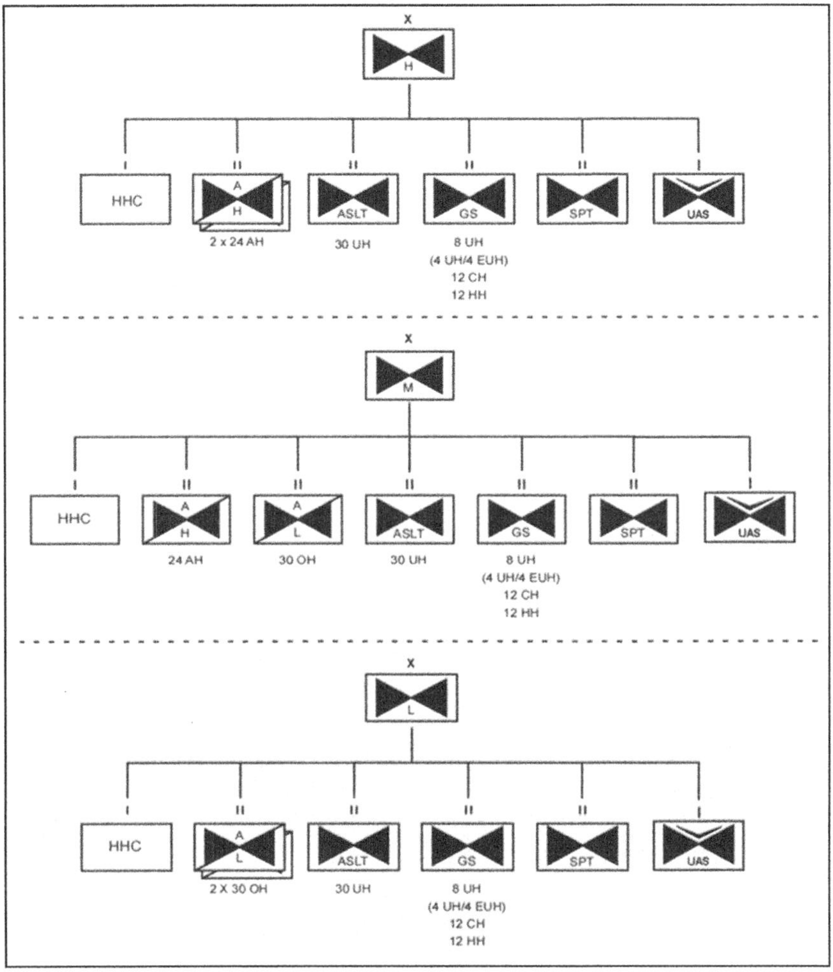

Figure 1-2. Heavy, medium, and light combat aviation brigade

Note. Medium aviation brigades also have a Pathfinder company located within the AHB.

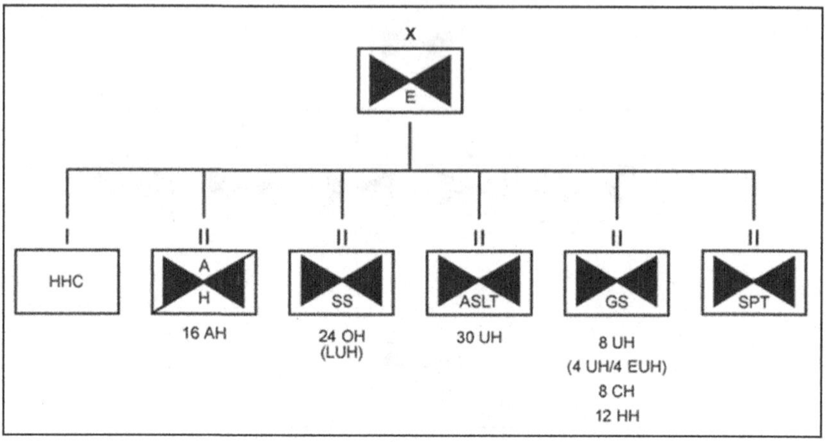

Figure 1-3. Combat aviation brigade (expeditionary)

THEATER AVIATION BRIGADES

1-24. Each theater aviation brigade (TAB) is structured to conduct assault or general support aviation missions in support of the theater and its subordinate commands.

1-25. Unlike CABs, TABs do not contain ARBs or ATS companies. It reinforces divisional aviation brigades with assault, general support, heavy lift, aeromedical evacuation, or fixed-wing (FW) assets..

1-26. TABs conduct, air assault, air movement, and sustainment operations for itself and to reinforce divisional aviation brigades.

1-27. The theater aviation brigade (figure 1-4, page 1-7) consists of an HHC, 3 AHBs with 30 UH-60s each; a GSAB (figure 1-5, page 1-7) with 4 UH-60s, 4 EUH-60s, 12 CH-47s, and 12 HH-60s; and an ASB. The theater aviation brigade (composite) consists of an HHC; 3 GSABs with 4 UH-60s, 4 EUH-60s, 12 CH-47s, and 12 HH-60s each; a FW battalion with 40 aircraft; and an ASB.

SUBORDINATE AVIATION BATTALIONS

1-28. The numbers and types of subordinate battalions included in a specific brigade are based on the brigade's mission. Separate companies or battalions may be assigned, attached, or placed operational control (OPCON) to aviation brigades. This situation presents challenges for C2 as the brigade staff must prepare plans and orders on the level of detail normally found at battalion level.

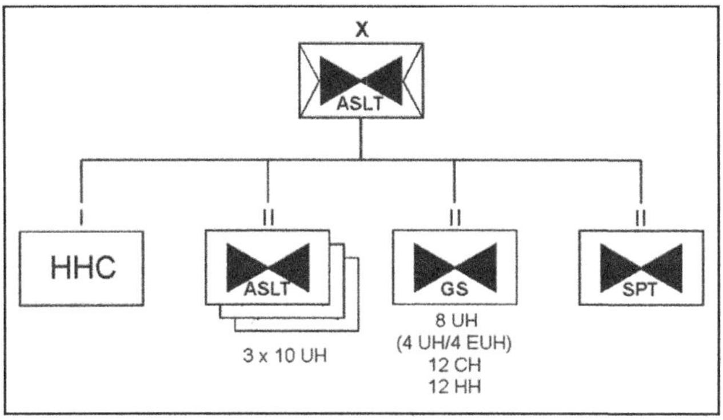

Figure 1-4. Theater aviation brigade

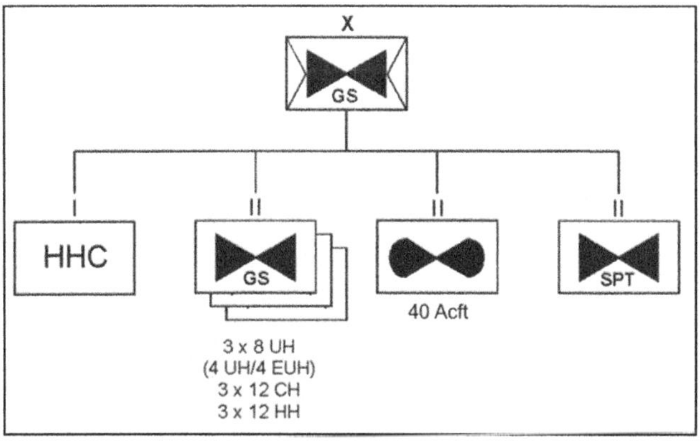

Figure 1-5. Theater aviation brigade (composite)

1-29. Subordinate battalions found in aviation brigades are—

- An ARS with 30 OH-58s (refer to FM 3-04.126).
- An ARB with 24 AH-64s (refer to FM 3-04.126).
- An assault helicopter battalion with 30 UH-60s (refer to FM 3-04.113).
- A GSAB with 4 EUH-60s, 4 UH-60s, 12 CH-47s, and 12 HH-60s (refer to FM 3-04.113).
- A fixed-wing aviation battalion with 40 FW aircraft.
- An S&S BN (currently equipped with 24 OH-58s, and in the future with the LUH).
- An ASB.
- A UAS company (refer to FMI 3-04.155).

1-30. The ARB/ARS can support early tactical operations (TACOPS) and ground maneuver with reconnaissance, security, and attack. It is effective against massed-moving, point (such as cave entrances, bunker apertures, and specific windows in buildings), and other hard or soft targets.

1-31. The AHB provides air assault and air movement capability.

1-32. GSAB assets participate in all brigade operations.

- The command aviation company (CAC) provides aerial C2 and GS.
- The heavy helicopter company (HvyHC) provides heavy-lift support.
- The air ambulance medical company (AAMC) provides aeromedical evacuation support.
- The ATS company provides ATS support.
- The UAS company provides intelligence, surveillance, and reconnaissance (ISR).

1-33. The fixed wing aviation battalion conducts flight missions required for battle command of theater operations; liaison among theater commands; and transportation of supplies, equipment, and personnel within the theater.

1-34. The security and support battalion is intended to support a variety of federal, state and homeland security missions.

1-35. The ASB plans, coordinates, and executes the aviation brigade's sustainment and signal support requirements.

1-36. UAS operations support battlefield commanders and their staffs as they plan, coordinate, and execute operations. UAS increase the situational awareness (SA) of commanders through ISR .

Chapter 2

Command and Control

The C2 system is an arrangement of personnel, information management (IM), procedures, equipment, and facilities essential to the commander in conducting operations (FM 6-0). There are two parts of this system; the commander and the control system. This chapter briefly discusses the brigade commander's role in C2 and details the purpose and characteristics of the CPs that form the brigade's control system. It also discusses the responsibilities and duties of staff officers, the brigade communications systems, and brigade-level planning to facilitate successful operations. Refer to FMI 5-0.1 for additional information on the C2 system.

SECTION I – OVERVIEW

2-1. Effective and efficient C2 is a process beginning and ending with the commander. The commander must develop techniques and procedures promoting an expeditious flow of information throughout the entire C2 process. These techniques and procedures should be in the unit's tactical standing operating procedures (TACSOP). FM 6-0 describes command and control.

2-2. C2 is the exercise of authority and direction by a properly designated commander over assigned and attached forces in the accomplishment of a mission (FMI 5-0.1). Command includes both authority and responsibility for effectively using available resources to accomplish missions.

Contents

COMMAND

2-3. Command is the art of motivating and directing people and organizations to accomplish missions. It requires visualization of the current state of friendly and enemy forces and the future state of those forces that must exist to accomplish the mission; and formulates concepts of operations to achieve victory.

CONTROL

2-4. Control is regulating forces and WFFs in executing the commander's intent. It helps commanders and staffs compute requirements, allocate means, and integrate efforts. Control is necessary in determining the status of organizational effectiveness, identifying variance from set standards, and correcting deviations from these standards. It permits commanders to acquire and apply means to accomplish their intent and develop specific instructions from general guidance. Ultimately, it provides commanders a means to measure, report, and correct performance. Control allows commanders freedom to operate, delegate authority, place themselves in the best position to lead, and synchronize actions throughout the operational area. Commanders exercise authority and direction through and with the assistance of a C2 system.

COMMAND AND CONTROL SYSTEM

2-5. C2 systems are defined as facilities, equipment, communications, procedures, and personnel essential to a commander for planning, directing, and controlling operations of assigned forces. The C2 system gives the commander a structure, a means to make and convey decisions, and the ability to evaluate the situation as it develops. Decisions and higher-level intent are then translated into productive actions.

2-6. Army battle command system (ABCS), as described in FM 6-0, provides unit commanders the electronic architecture to build SA. Signal planning increases the commander's options by providing requisite signal support systems for varying operational tempos. These systems pass critical information at decisive times; thus, they leverage and exploit tactical success and make future operations easier. Appendix B contains additional information on ABCS. ABCS consists of—

- Global Command and Control System-Army (GCCS-A).
- Army Tactical Command and Control System (ATCCS).
- Force XXI Battle Command Brigade and Below (FBCB2).

2-7. The airspace command and control (AC2) system is the airspace management component of the Army air-ground system. It outlines the Army's integration of airspace usage and C2 within the framework of the theater air-ground system (TAGS). Ground and air joint forces must be able to operate aerial vehicles and weapons systems within shared airspace with maximum freedom consistent with priorities, the degree of operationally acceptable risk, and the joint force commander's (JFC's) intent. Refer to FM 3-52 for additional information.

BATTLE COMMAND

2-8. Battle command is the art and science of visualizing, describing, directing, leading forces, and assessing forces in operations against a hostile, thinking, and adaptive enemy. Battle command applies leadership to translate decisions into actions—by synchronizing forces and WFFs in time, space, and purpose—to accomplish missions. Battle command is guided by professional judgment gained from experience, knowledge, education, intelligence, and intuition. Refer to FM 3-0 for additional information.

2-9. Commanders visualize the operational environment (OE), the desired end state, and a broad concept of how to transform the conditions of the environment from the current state to the end state. Visualization is passed to the staff and subordinate commanders through commander's intent and planning guidance; commanders express gaps in their operational knowledge through commander's critical information requirements (CCIRs). Commanders direct actions to achieve success and lead forces to mission accomplishment. Refer to FM 3-0 for additional information.

BATTLE COMMAND ON THE MOVE

2-10. BCOTM is the ability to lead Soldiers and command all elements of combat power by shaping and sustaining decisive actions seamlessly while on the battlefield. Commanders must synchronize the elements in close combat from any vantage point on the battlefield. The ability to reposition rapidly as the situation develops enables commanders to better see the battlefield and be at the critical point at the critical time.

2-11. BCOTM performs the following functions:

- Sees and understands the common operational picture (COP).
- Directs and controls maneuver operations.
- Controls direct/indirect fires and effects.
- Monitors enemy and intelligence activities.
- Synchronizes forces.
- Directs reconnaissance/counter reconnaissance operations.
- Executes operation orders (OPORDs).
- Issues fragmentary orders (FRAGOs).
- Receives and renders reports.

SECTION II – BRIGADE COMMAND GROUP

2-12. The brigade command group consists of the commander and selected staff members who accompany commanders and enable them to exercise command and control away from the command post, (FMI 5-0.1) The command group is organized and equipped to suit the commander's decisionmaking and leadership requirements. The members typically consist of at least the executive officer and the primary staff members. This group operates as dictated by the commander and at any location that provides for effective C2.

COMMANDER

2-13. The brigade commander commands and controls the aviation brigade. He or she is responsible for the outcome of his or her force's actions. The variety and impact of tasks confronting him or her are unique. His or her main concerns are accomplishing the mission and ensuring the welfare of Soldiers.

2-14. The brigade commander visualizes the operating environment, analyzes the situation ,defines the mission, and directs actions for the mission's execution. The brigade commander controls the current operation and provides guidance for planning future operations.

2-15. The aviation brigade's forces influence an enormous area of the OE therefore; the commander must have extensive situational awareness of the environment. Tactical decisions must be constantly aimed at synchronizing his or her unit's combat efforts with those of multiple combined, joint, interagency, and multinational forces.

EXECUTIVE OFFICER

2-16. The executive officer (XO) is second in command and principal assistant to the commander. The scope of XO duties is often tailored by the commander's requirements. Normally, the XO directs, supervises, and ensures coordination of staff work and logistics, except in those specific areas reserved by the brigade commander. He or she must understand the commander's guidance and intent as he or she supervises the entire process. He or she ensures the staff has the information, guidance from the commander, and facilities it needs. He or she determines timelines for the staff, establishes backbrief times and locations, enforces the IM plan, and provides any unique instructions to guide the staff in completing the military decisionmaking process (MDMP).

2-17. The XO is the primary senior leader on the brigade staff and additionally ensures appropriate planning and execution of the logistics and maintenance effort.

STAFF ORGANIZATION

2-18. The staff consists of officers and enlisted personnel who plan, supervise, and synchronize planning and execution of aviation brigade operations, to include sustainment, according to the brigade commander's concept and intent (figure 2-1, page 2-4). Except in scope, duties and responsibilities of the brigade staff are similar to those of a BCT staff and are often similar to the higher echelon staff. Key staff personnel must be positioned on the battlefield where they can carry out their duties.

STAFF RESPONSIBILITIES

Standing Operating Procedures

2-19. The aviation brigade SOP must clearly define the responsibilities of key personnel to preclude conflicts and ensure all functions are planned, coordinated, executed and supervised. SOPs streamline procedures and reports by providing standardized processes, briefings and reports. The SOP also identifies specific individuals and sections, who request, receive, process, and disseminate information.

2-20.

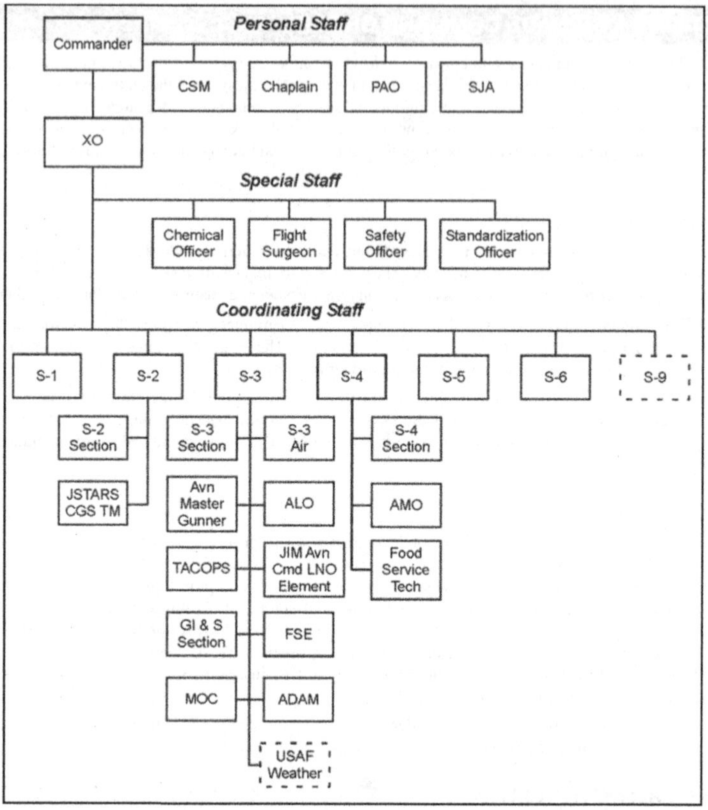

Figure 2-1. Aviation brigade staff organization

Running Estimates

2-21. Running estimates may be informal at brigade level and below; however, they must address battlefield activity, project course of action (COA), and predict results. Careful IPB, selection of the most important enemy indicators, and development of contingency plans facilitate estimates and allow timely response. The key person in this process is the XO. He or she ensures the staff maintains a proper perspective. Refer to FMI 5-0.1 for additional information on running estimates.

Communications with the Commander

2-22. The staff must provide the commander with critical, concise, accurate information, especially that information the commander identified as CCIR. Critical information is communicated to the commander on a priority basis set by the commander or the XO's guidance.

2-23. Information flow, both horizontally and vertically, must be on a priority basis. Operational conditions dictate priorities.

PERSONAL STAFF

2-24. The personal staff works under the commander's immediate directon, but may also work through the XO or a specific coordinating staff officer for coordination and control purposes. Members of the personal staff normally have direct access to the commander due to the nature of their assigned duties.

COMMAND SERGEANT MAJOR

2-25. The command sergeant major (CSM) acts in the name of the commander and is his or her primary advisor concerning enlisted soldiers. The CSM focuses attention on functions critical to the success of the operation. The CSM assists the commander in the following ways:

- Monitors noncommissioned officer (NCO) development, promotions, and assignments.
- Identifies, plans, and assesses soldier training tasks to support the performance of collective (unit) tasks on the mission essential task list (METL).
- Monitors subordinate unit morale.
- Provides recommendations and expedites procurement and preparation of enlisted replacements for subordinate units.
- Monitors food service and other logistics operations.
- Conducts informal investigations.
- Assists in controlling brigade movements.
- May lead the brigade advance or quartering party during a major movement, coordinating closely with the HHC commander.

CHAPLAIN

2-26. The aviation brigade chaplain provides unit level religious support to all personnel assigned/attached to the brigade including nondenominational coverage and ministry for mass casualties and hospitalized members of the brigade. The chaplain advises the commander on religious, moral, and Soldier welfare and morale issues. He or she establishes liaison with unit ministry teams (UMTs) of higher and adjacent units. He or she also supervises subordinate unit chaplains and provides backup services as required. The chaplain and chaplain's assistant compose the UMT, usually operating from the same location as the adjutant.

PUBLIC AFFAIRS OFFICER

2-27. The public affairs officer (PAO) leads the public affairs cell, which coordinates media and community relations. The public affairs cell provides coordination of public affairs and strategic messaging advice and counsel to the commander. The public affairs cell further coordinates requests for public affairs support from the division or higher echelon command, and provides support to all accredited media operating in the aviation brigade area of operations (AO).

STAFF JUDGE ADVOCATE

2-28. The brigade judge advocate, along with the operational law judge advocate and paralegal NCO, form the brigade operational law team (BOLT). The brigade judge advocate serves as a personal and special staff officer. The BOLT provides legal advice to the commander and staff. The members of the BOLT serve as subject matter experts (SMEs) on ROE, targeting, international law, law of armed conflict (including treatment of detainees, enemy prisoners of war, civilians on the battlefield and other noncombatants), and all other legal aspects of operations. The BOLT provides or coordinates with other legal offices for legal services in military justice, administrative and civil law, contract and fiscal law, claims, and legal assistance. The paralegal NCO provides administrative and paralegal support to the judge advocates in the BOLT and supervises the paralegals in the aviation battalions.

SPECIAL STAFF

2-29. Special staff officers help the brigade commander and other staff members perform their functional responsibilities.

CHEMICAL OFFICER

2-30. The chemical officer operates from the operations staff officer (S-3) section and advises the commander on—

- Chemical, biological, radiological, and nuclear (CBRN) defense.
- Decontamination.
- Smoke operations.
- Use of obscurants and flame.

FLIGHT SURGEON

2-31. The brigade flight surgeon advises and assists commanders on matters concerning the medical condition of the command including preventive, curative, and restorative care. The brigade flight surgeon, with subordinate unit flight surgeons, conducts flight physicals for unit personnel. The flight surgeon determines requirements for the requisition, procurement, storage, maintenance, distribution, management, and documentation of medical equipment and supplies for the brigade HHC. The flight surgeon, in conjunction with the physician assistant, operates the brigade aid station normally located in the brigade assembly area (AA).

SAFETY OFFICER

2-32. The safety officer (SO) serves as the principle advisor to the commander and staff on aviation and ground safety matters. The SO assists the staff and commander during the composite risk management (CRM) process, and monitors brigade and subordinate unit missions to identify and address potential hazards. He or she recommends actions to ensure safe mission accomplishment.

2-33. The brigade SO implements and monitors safety-related programs and advises subordinate unit SOs as required. The brigade SO is responsible for standardization of the safety contents of the reading files. The SO should be rated in the highest-density type aircraft in the brigade.

STANDARDIZATION INSTRUCTOR PILOT

2-34. The SP is the primary advisor to the commander for the flight standardization program. The SP develops, integrates, implements, monitors, and manages the aircrew training and standardization programs. Along with the Master Gunner and TACOPS officer, the brigade SP provides advice and counsel to commanders on mission planning and aviation TTP. The brigade SP provides advice on the crew selection process, and in concert with the Master Gunner, monitors the employment of aircraft systems, sensors, and weapons. He or she is a principal trainer and peer leader for subordinate unit SPs and instructor pilots (IPs). The brigade SP acts as the coordinating staff officer for reading file standardization. Additionally, the SP assists the commander in the following ways:

- Provides recommendations and assists in expediting warrant officer replacements for subordinate units.
- Monitors warrant officer development and assignments.
- Conducts informal investigations.
- Performs specific missions as directed by the brigade commander.

COORDINATING STAFF

2-35. The coordinating staff is composed of the commander's principle assistants who are responsible for one or a combination of broad fields of interest (personnel, intelligence, operations, logistics, planning, and communications). Coordinating staff members help the commander coordinate and supervise execution of plans, operations, and activities. Collectively, through the XO, they are accountable for the commander's entire field of responsibility.

HUMAN RESOURCE SECTION

2-36. The personnel staff officer (S-1) has coordinating responsibility for finance, religious activities, public affairs, and legal services support of the unit. The S-1 is normally collocated with the S-4 in the sustainment cell of the main CP. The S-1 and S-4 will cross-train to enable continuous operations.

2-37. The S-1 is responsible for all matters concerning human resources including personnel readiness and services. The S-1 manages personnel strength and replacement; works with the flight surgeon to plan health services; coordinates morale support activities and legal, financial, and postal services; maintains the awards program; oversees administration of discipline, law, and order with the provost marshal (if present) and brigade judge advocate; and provides casualty operations management.

INTELLIGENCE SECTION

Intelligence Officer

2-38. The intelligence staff officer (S-2) leads the intelligence staff section consisting of the S-2 section, tactical command post (TAC CP), and joint surveillance target attack radar system (JSTARS) common ground station (CGS) team. The S-2 is responsible for all matters concerning ISR. The S-2 provides current information and analyzed intelligence of tactical value concerning terrain, weather, and the enemy.

Intelligence Section

2-39. The S-2 section provides combat intelligence, including collecting and processing information. The S-2 section prepares intelligence collection plans; receives and analyzes battlefield information; disseminates intelligence products; and provides up-to-date intelligence information to assist in planning for and coordinating operations.

2-40. The S-2 section performs the following functions:

- Coordinates intelligence activities.
- Converts information requirements of the commander into the CCIR.
- Facilitates the IPB process.
- Helps develop the decision support template.
- Develops the intelligence collection and reconnaissance and surveillance plan.
- Frequently updates the commander and staff on the enemy situation.
- Maintains isolated personnel reports.
- Performs intelligence support to effects and targeting.
- Processes relevant information to assist in creating the COP.
- Provides aircrews with intelligence summaries and pre-flight update briefs.
- Conducts post-flight debrief for intelligence gathering.
- Identifies potential high-payoff targets (HPTs) and high-value targets (HVTs).

Joint Surveillance Target Attack Radar System Common Ground Station Team

2-41. The JSTARS CGS team provides the commander with near-real-time wide area surveillance and deep targeting data on moving and fixed targets during daylight and darkness in all weather conditions. The team receives, processes, correlates, and disseminates imagery data from the JSTARS. The CGS receives, manipulates, displays, stores, and disseminates JSTARS, unmanned aircraft system (UAS), Army aviation, signal intelligence, broadcast intelligence and secondary imagery from tactical, theater, and national systems.

Operations Section

Operations Staff Officer

2-42. The S-3 is responsible for matters pertaining to the organization, employment, training, and operations of the brigade and supporting elements. The S-3 monitors the battle, ensures necessary sustainment assets are provided when and where required, anticipates developing situations, and executes the appropriate actions within the scope of the commander's guidance.

Operations Section

2-43. The S-3 section provides planning and task organization of brigade elements for operations. The S-3 section maintains routine reporting and coordinates activities of liaison personnel. It also coordinates consistently with the plans staff officer (S-5) to synchronize future operations and the transition from current operations to a future operation. The S-3 section, through the command, control, communications, and computer operations (C4OPS) cell, ensures procedures are in place ensure effective communications. The S-3 section maintains close coordination with the S-4 and S-1 for brigade logistics and personnel status.

S-3 Air

2-44. The aviation brigade S-3 Air is responsible for coordinating air space issues, JAAT operations, and joint air operations. The aviation brigade S-3 Air is responsible for the integration of A2C2 planning, coordination, and airspace deconfliction for combined arms JIM operations in the aviation brigade.

2-45. The S-3 Air oversees the following functions:

- Developing AC2 procedures, plans, SOPs, and annexes.
- Submitting requests for airspace coordinating measures (ACMs).
- Ensuring AC2 restrictions are incorporated in FS planning.
- Obtaining and distributing the current air control order (ACO) for each subordinate battalion size headquarters.
- Obtaining and distributing applicable portions of the special instructions (SPINS) and air tasking order (ATO) to subordinate units.
- Incorporating applicable AC2 measures into the aviation brigade scheme of maneuver.
- Maintaining the AC2 overlay.
- Establishing and monitoring the flight following net (ATS network) for brigade aircraft, when required.
- Assisting the S-3 and fire support element (FSE) in joint suppression of enemy air defense (J-SEAD) fires planning.
- Coordinating for additional Army and joint aviation support to support aviation operations such as movement of unit equipment, supplies, ammunition, and fuel.
- Assisting the S-3 in planning, organizing, and coordinating aviation brigade participation in JAAT operations.

Tactical Operations Officer

2-46. TACOPS officers advise the commander and staff on appropriate aircraft survivability equipment (ASE) techniques and procedures and integration of joint assets for each mission. TACOPS officers conduct the ASE part of the CRM process, integrate the unit's operation plan (OPLAN) into the theater airspace structure, assist with development of unit TTP, manage the organization's PR program, and are involved in the MDMP, close air support (CAS), and FS. TACOPS officers are primary SMEs for the organic AMPS and its associated products. They also assist in training members of ground maneuver brigade aviation elements (BAEs) and subordinate unit TACOPS officers.

Air Liaison Officer

2-47. Depending on the expected types of missions, an air liaison officer (ALO) may be provided. The ALO is a USAF officer who is a member of the tactical air control party (TACP). The ALO may serve as a forward air controller (FAC) or have additional officers assigned to the ALO as FACs. The ALO advises the commander and staff on employment of air support, including CAS, air interdiction (AI), J-SEAD, aerial reconnaissance, and airlift. In the absence of an ALO, the S-3 ensures these duties are accomplished.

Aviation Master Gunner

2-48. The master gunner serves as the advisor to the commander and staff on aviation weapons employment. He or she interfaces with the S-3, S-5 and TACOPS officer during the mission planning process and is primarily responsible for the targeteering and weaponeering process. Master gunners develop and recommend tactics, techniques, and procedures (TTP) to optimize aviation weapons effectiveness and collaborate with the TACOPS officer on route selection.

2-49. The brigade master gunner is a primary advisor to the S-3 assisting in forecasting and allocating ammunition and monitoring gunnery training device usage. The master gunner monitors gunnery-related programs throughout the brigade, and advises subordinate unit master gunners as required. He or she supports the battalions in developing advanced gunnery training tables to include realistic target arrays, and assists in the coordination of scheduling with local range-control officials. The master gunner should be SP-qualified and current in one of the attack/reconnaissance aircraft within the brigade.

Joint, Interagency, and Multinational Aviation Command Liaison Element

2-50. The JIM aviation command liaison element provides necessary liaison between the aviation brigade and JIM aviation CP. This element is essential for planning and execution of aviation missions in the brigade area, and represents the brigade commander at a separate location to facilitate communication and aviation planning.

2-51. Brigade commanders must empower liaison teams to act on their behalf and ensure they are fully supported. In return, commanders expect positive two-way communication. Liaison teams do not commit aviation brigade assets or approve changes to a plan without brigade S-3 or commander coordination.

2-52. Liaison teams must have access to current brigade status information to provide the most accurate picture of capabilities. Constant communication with the parent unit is essential for updates on aircraft availability, maintenance, and FARP status.

Fire Support Element

2-53. The FSE provides FS planning, coordination, and execution through the following tasks:

- Conducts nonlethal FS operations and information operations (IO).
- Conducts lethal FS.
- Employs fire support coordinating measures (FSCMs).
- Provides firepower in support of operational maneuver.
- Provides CAS integration for surface forces.
- Employs positive and procedural control measures.

Fire Support Coordinator

2-54. The FSE is headed by the fire support coordinator (FSCOORD). He or she provides support for the scheme of maneuver with fires and direct targeting process execution (detect, deliver, and assess). Both missions are critical to the success of aviation operations. The FSCOORD accomplishes this by close coordination with the S-3 and brigade commander.

2-55. The FSCOORD plans, controls, and synchronizes all lethal and nonlethal FS for brigade operations. He or she coordinates J-SEAD, and integrates and coordinates offensive IO into FS planning. He or she works with the S-3 and A2C2 element regarding field artillery (FA) firing unit locations and changes to FSCMs and ACMs. The FSCOORD maintains communications with supporting artillery.

Geographic Information and Services Topographic Engineer Element

2-56. The geospatial information and services topographic engineer element provides the following support:

- Database management.
- Topographic database development.
- Terrain analysis.
- Topographic survey.
- Topographic production.

Air Defense and Airspace Management Cell

2-57. The air defense and airspace management (ADAM) cell provides liaison between the aviation brigade and air defense (AD) units. The employment of the ADAM element is essential in deconflicting airspace and preventing fratricide of friendly aircraft operating throughout the AO.

United States Air Force Weather Team

2-58. The brigade may receive weather team support when local facilities or assets are unable to provide required weather information. The team locates with the brigade main CP and provides their own weather equipment.

Medical Operations Cell

2-59. The medical operations cell (MOC) provides assistance in planning and coordination for air ambulance employment and utilization. The MOC consists of a medical service corps officer and NCO who provide assistance with synchronization of the air and ground evacuation plan. It is the conduit for communication between higher echelons of medical C2 and the CAB. The medical service corps officer and NCO also manage medical treatment facility (MTF) information from medical support commands and surgeon cells of higher echelons including hospital locations and status, evacuation routes, casualty collection points, and ambulance exchange points. The MOC also performs the following functions:

- Establishes flight procedures specific to aeromedical evacuation missions within the CAB. This may include special routes or corridors as well as procedures for escort aircraft link-up.
- Ensures lines of communication (LOCs) to supported units and higher echelons of medical command are available. The MOC also ensures supported units understand aeromedical evacuation procedures and capabilities—an educational endeavor that is an ongoing process.
- Establishes aeromedical evacuation briefing and launch procedures. Ensures 24-hour access to those able to launch high and very-high risk missions.
- Maintains awareness of the tactical and medical situation. Coordinates with medical regulators at higher echelons to efficiently conduct GS and works in concert with adjacent units.
- Assists the AAMC and GSAB/CAB staff in conducting aeromedical evacuation operations.

LOGISTICS SECTION

Logistics Staff Officer

2-60. The logistics staff officer (S-4) is responsible for brigade level coordination of all external and internal logistics support including supply, maintenance, transportation, and equipment status records. The S-4, as the brigade's logistics planner, coordinates with battalion S-4s, separate company supply officers, or first sergeants (1SGs) regarding the status of maintenance, equipment, and supplies. The S-4 ensures logistics visibility for the brigade commander and staff. The S-4 develops the logistics support plan, and coordinates with supporting units and higher headquarters staffs ensuring logistics support is continuous.

Logistics Section

2-61. The S-4 section provides supervision and coordination of food service, supply, transportation, and maintenance support for the brigade. S-4 section responsibilities include–

- Recommending basic loads and supply requirements.
- Recommending the ammunition required supply rate (RSR) to the S-3.
- Coordinating all classes of supply (except class VIII).
- Coordinating equipment recovery, evacuation, and repair.
- Conducting planning for operational movement control and mode and terminal operations.
- Coordinating with the civil affairs cell for host nation support.
- Coordinating services including water purification, mortuary affairs, aerial resupply, laundry, shower, and food preparation.
- Coordinating battlefield procurement and contracting.

Brigade Aviation Materiel Officer

2-62. The brigade aviation materiel officer (BAMO) is an aviation staff officer assigned to the S-4 section. The BAMO is an advisor to the brigade commander and staff for aviation maintenance issues. The aviation materiel officer (AMO) ensures close coordination with the ASB on aviation maintenance issues. The brigade AMO is a trainer and peer leader for the subordinate unit AMOs. He or she should be rated in the highest-density type aircraft in the brigade. See FM 3-04.500.

2-63. The BAMO is the primary adviser to the brigade commander for the effectiveness and efficiency of the aviation maintenance program in the unit. The BAMO recommends actions and forecasts future capabilities based on the existing maintenance posture.

2-64. The BAMO plans maintenance actions based on operational necessities. He or she also maintains a daily status of all aircraft in the combat aviation brigade. The BAMO is normally a maintenance evaluator, responsible for the following:

- Provides advice to the battalion commander on all aviation maintenance and logistics issues.
- Oversees quality control functions.
- Resolves aircraft maintenance standardization issues.
- Attends the brigade safety and standardization meeting.

2-65. The BAMO and battalion AMOs should work hand-in-hand with AMC production control personnel.

2-66. The BAMO manages personnel, supply, equipment, and facility assets to maintain and repair Army rotary- and fixed-wing aircraft. He or she organizes maintenance elements to inspect, service, test, disassemble, repair, reassemble, adjust, and replace parts and retest aircraft or aircraft components. He or she prepares, implements, and maintains standing operating procedures for management of maintenance activities.

2-67. The BAMO interprets regulations; technical manuals, including ETMs/IETMs; and orders pertaining to maintenance and logistics actions of Army aircraft for commanders and subordinates. He or she supervises the aviation equipment maintenance and repair shop, section, or platoon. He or she also directs maintenance and accountability of organizational test equipment, supplies, and recovery equipment.

2-68. Enlisted personnel may be assigned to the section to assist the BAMO in order to provide continuous maintenance and logistics information to the commander and staff. The BAMO keeps the command informed about current and future capabilities based on the current maintenance posture and plans maintenance actions based on operational necessities.

Food Service Technician

2-69. The aviation food service officer is a staff officer assigned to the S-4 section. He or she is an advisor to the brigade commander and staff for the food service program and class I issues. The food service officer evaluates field feeding requirements and develops milestone plans to support major field exercises.

He or she reviews and monitors requisitions for classes I, III, and IX supplies, and coordinates planning of food service support for field training. The food service officer evaluates garrison and field feeding operations, ensuring food service personnel are complying with food service regulations relative to food preparation, service, accountability, and sanitation.

PLANS SECTION

Plans Staff Officer

2-70. The S-5 is responsible for planning future operations. The S-5 monitors the COP, stays abreast of current operations by coordinating with the S-3, and plans sequels accordingly. The S-5 plans for operations to be conducted in the next phase, normally occurring in the brigade's contingency or orientation planning horizons (refer to FM 5-0 for additional information). However, the commander may task him or her to plan operations in the current phase or near-term horizon. Other S-5 responsibilities include—

- Producing OPORDs and warning orders (WARNOs) transition to future operations.
- Coordinating with the S-3 for transitioning from current to future operations.
- Producing current OPLANs for the S-3, when requested.
- Participating in the targeting process.
- Performing long-range assessment of an operation's progress.

COMMUNICATIONS-ELECTRONICS SECTION

Signal Officer

2-71. The S-6 advises the commander on signal matters, CP location, and signal facilities, assets, and activities. He or she maintains authority and responsibility for all network operations within the aviation brigade. Execution of network operations is the responsibility of the network support company (NSC) within the ASB.

2-72. The S-6 and NSC commander operate in close communication, resulting in a unity of effort for communications support to the aviation brigade. The NSC commander reports all network-associated issues to the S-6. The aviation brigade S-6 and his or her staff plan Command, Control, Communications and Computer Operations (C4OPS) support for the aviation brigade CPs and subordinate units organic to, assigned to, or operating within the aviation brigade AO. The aviation brigade S-6 works closely with both the division Assistant Chief of Staff- C4OPS (G-6) and the NSC commander.

Communication-Electronics Section

2-73. The S-6 section plans, coordinates, and oversees implementation of communications systems. It performs unit-level maintenance on ground radio and field wire communications equipment. It installs, operates, and maintains the radio retransmission site. This section monitors the maintenance status of signal equipment, coordinates preparation and distribution of the signal operation instructions (SOI), and manages communications security (COMSEC) activities. The S-6 section's responsibilities include supervision of electronic mail on both unclassified and classified nets and the local area network (LAN).

CIVIL-MILITARY OPERATIONS SECTION

Civil-Affairs Staff Officer

2-74. A civil affairs staff officer (S-9) is normally not available to the brigade. However, in certain operations, one may be designated or attached. The S-3 is responsible for S-9 when no S-9 is provided. In operations where the areas of responsibility for the S-3 and S-9 overlap, the S-9 is subordinate to the S-3. S-9 personnel working in any of the brigade's subordinate unit areas are subordinate to the commander of that subordinate unit, regardless of rank.

Civil-Military Operations Center

2-75. When accomplishing S-9 duties, the designated officer may have to coordinate with a civil-military operations (CMO) center. The CMO center is formed from civil affairs assets and serves as the primary interface between the U.S. Armed Forces and local civilian population, humanitarian organizations, nongovernmental organizations, private volunteer organizations, other international agencies, multinational military forces, and other agencies of the U.S. government. The CMO center ensures continuous coordination among key participants regarding civil-military matters. It is a flexible, mission-dependent organization formed at brigade and higher-level headquarters.

SECTION III – BRIGADE AVIATION ELEMENT

2-76. The BAE is in the light, medium, and heavy BCTs and the battlefield surveillance brigade. Though not part of the aviation brigade staff; it plays an important role in aviation planning, execution, and AC2. Aviation brigade and aviation TF commanders must be personally involved in training, equipping, and mentoring BAEs.

2-77. The BAE is a planning and coordination cell organic to each BCT whose major function is incorporating aviation into the ground commander's scheme of maneuver. The BAE focuses on providing employment advice and initial planning for aviation missions including; UAS, airspace planning and coordination, and Army aviation synchronization with the ALO and FSCOORD. The BAE also coordinates directly with the aviation brigade or supporting aviation TF for detailed mission planning.

2-78. The BAE does not take the place of aviation TF involvement in the planning process. It assists the BCT in aviation planning and provides the aviation brigade or supporting aviation TF leadership with BCT mission information. It is critical for aviation commanders and S-3s to participate and lead aviation mission planning in support of the BCT.

2-79. The BAE is organized and equipped to support the BCT and consists of a sufficient number of personnel for 24-hour operations. It uses the ABCS, which can network with the joint planning and communications architecture. The BAE is normally composed of a major, captain, senior warrant officer, and three enlisted Soldiers. See training circular (TC) 1-400 for additional BAE information.

2-80. The BAE is involved in the mission from receipt of the WARNO from higher headquarters through planning; movement to the port of embarkation (POE); deployment; reception, staging, onward movement, and integration (RSOI) into the force; MDMP; combat operations; and redeployment, reintegration, reconstitution, and retraining (R4).

2-81. The BAE provides—

- Integration and synchronization of aviation into the BCT commander's scheme of maneuver.
- AC2 planning, coordination, airspace deconfliction for combined arms, and JIM operations.
- Focus on incorporating aviation into the commander's plan.
- Direct coordination with aviation brigades and ABTFs.
- Close integration/synchronization with the ALO and FSCOORD.

SECTION IV - AVIATION LIAISON TEAMS

2-82. The aviation brigade headquarters contains an aviation liaison team to facilitate operations with higher headquarters and/or supported ground maneuver units. Since the aviation brigade has limited liaison assets available, liaison teams are organic to aviation battalions and represent their units as directed to facilitate air-ground integration and planning.

2-83. Although a BAE conducts many of the functions traditionally performed by liaison officers (LNOs), aviation liaison teams remain a critical part of the process and thus must be staffed appropriately.

2-84. While a BAE works directly for the BCT commander as a permanent member of his or her staff, aviation liaison teams represent the supporting aviation TF at designated maneuver headquarters for the duration of a specific operation.

2-85. If collocated with a BAE, the liaison team normally works directly with the brigade aviation officer as a functioning addition to the BAE staff section. Effective employment of LNOs is imperative for coordination and synchronization. Often aviation liaison teams coordinate with the BAE and proceed to a supported ground maneuver battalion location. An example would be an aviation liaison team in support of an infantry battalion performing an air assault to seize a key piece of terrain.

2-86. Liaison teams maintain and provide current—

- Aviation unit locations.
- Aircraft/equipment status.
- Crew availability and fighter management cycle status.
- Class III/V status.
- Continuous updates to the aviation commander and staff on the BCT's plan.

SECTION V – AVIATION BRIGADE COMMAND POSTS

2-87. To assist in controlling operations, the brigade commander organizes his or her staff into CPs that provide staff expertise, communications and information systems that work in concert to aid the commander in planning and controlling operations. (Refer to FM I 5-0.1 and Appendix G of this publication) Activities common to all CPs include—

- Maintaining running estimates and the common operating picture.
- Information management (See FM 3-0 and FM 6-0).
- Develope and disseminate orders (see FM 5-0)
- Controlling operations (FMI 5-0.1).
- Assesing operations (FMI 5-0.1).
- Coordinating with higher, lower, and adjacent units
- CP administration

2-88. CPs throughout the brigade serve the C2 needs of the commander and staff. The dynamics of the battlefield require the highest level of organizational and operational efficiency within every CP. C2 facilities include—

- Main CP.
- TAC CP.
- Command group.
- Subordinate unit CPs.

MAIN COMMAND POST

2-89. The main CP is a C2 facility containing the portion of unit headquarters in which the majority of planning, analysis, and coordination occurs. The main CP is organized into seven functional cells as described below. The brigade XO is responsible for the main CP. (See FMI 5-0.1) See appendix G for a sample layout of the brigade CP.

2-90. During combat operations the commander and staff are normally located in the brigade main CP. Often the commander and selected members of the main CP will go forward as a tailored force to C2 current operations. This may be the TAC CP or other location.

ELEMENTS

2-91. The main CP is composed of seven cells, orienting loosely on the WFFs—

- Current Operations (C2)
- Movement and Maneuver/Protection
- Fire Support/Protection
- Intelligence

- Sustainment
- Plans
- C4OPS

2-92. The main CP is the primary C2 structure for the brigade. Its primary missions are to control operations and to prepare and publish orders and plans. The commander operates from the main CP when not operating from the TAC CP, command vehicle, or an aircraft. The cells usually operate in shifts ensuring 24-hour capability. The plans cell may or may not operate on a 24-hour cycle.

Current Operations

2-93. The current operations cell is where the commander usually locates when at the main CP. The battle captain, usually the most experienced operations officer other than the S-3, continuously monitors operations within the operations cell ensuring proper personnel are available for the mission at hand. He or she does not command the battle, but performs battle tracking and makes operational decisions within assigned responsibilities.

2-94. The current operations cell is responsible for assessing the current situation while regulating forces and WFFs in accordance with the commander's intent. Normally, all staff sections are represented in the current operations cell. The unit's S-3 supported or an assistant S-3, leads this cell. Members of the movement and maneuver cell are normally also members of the current operations cell.

2-95. Staff representatives in the current operations cell actively assist subordinate units. They provide them information, synchronize their activities, and coordinate their support requests. The current operations cell solves problems and acts within the authority delegated by the commander. It also performs short-range planning using the MDMP in a time-constrained environment or makes decisions and resynchronizes operations.

Fire Support/Protection

2-96. The fire support cell coordinates activities and systems that provide collective and coordinated use of Army indirect fires and joint fires. This includes tasks associated with targeting and the targeting process. The fire support cell integrates lethal and nonlethal fires, including offensive IO, through the targeting process. The FSCOORD leads this cell.

Intelligence

2-97. The intelligence cell coordinates activities and systems that facilitate understanding the enemy, terrain, weather, and civil considerations. This includes tasks associated with intelligence preparation of the battlefield and ISR. The unit's S-2 leads this cell.

Protection

2-98. The protection cell may be physically divided among the movement and maneuver element and the fire support element. It coordinates the activities and systems that preserve the force. This includes protecting personnel, physical assets, and information. Commanders normally select this cell's leader from among the air and missile defense coordinator, chemical officer, engineer coordinator, and provost marshal.

Movement and Maneuver/Protection

2-99. The movement and maneuver cell coordinates activities and systems that move forces to achieve a position of advantage in relation to the enemy. This includes tasks associated with employing forces in combination with direct fire or fire potential (maneuver), force projection (movement), mobility, and countermobility. The movement and maneuver cell may also form the base of the current operations cell. The unit's G-3/S-3 or a deputy G-3/S-3 leads this cell.

Plans Cell

2-100. The plans cell is led by the S-5 and consists of personnel required to plan for operations, such as representatives from the S-2, FSE, ALO, S-1, S-4, S-6, engineer, S-9, and attached units. The plans cell is responsible for planning operations for the mid- to long-range time horizons. It develops plans, orders, branches, and sequels. This cell is also responsible for long-range assessment of an operation's progress.

Sustainment

2-101. The sustainment cell is normally led by the S-1 and S-4 while supervised by the XO. The sustainment cell coordinates activities and systems that provide support and services to ensure freedom of action, extend operational reach, and prolong endurance. It includes the tasks listed in paragraph 1-29. The commander normally designates the S-1 or S-4 as this cell's leader.

Command, Control, Communications, and Computers Operations

2-102. The C4OPS cell coordinates activities and systems that provide support to continuous and assured communications. This includes tasks associated with C4OPS, network operations, and information systems support to information management. The S-6 leads this cell.

Main Command Post Functions

2-103. The main CP coordinates, directs, and controls operations and plans for future operations. The main CP—

- Communicates with subordinate, higher, and adjacent units.
- Informs and assists the commander and subordinate commanders.
- Operates on a 24-hour basis.
- Conducts future planning continuously.
- Estimates the situation continuously.
- Maintains SA across the Army WFFs.
- Receives, evaluates, and processes tactical information from subordinate units and higher headquarters.
- Maintains maps graphically depicting friendly, enemy, and noncombatant situations.
- Maintains journals.
- Validates and evaluates intelligence.
- Controls all immediate FS including CAS for units under aviation brigade C2 (may also be done by TAC CP).
- Coordinates airspace C2 and AD operations.
- Coordinates maneuver and sustainment requirements.
- Coordinates terrain management for C2 facilities.
- Coordinates and tracks sustainment (logistics, air and ground maintenance capabilities, and status).
- Makes recommendations to the commander.
- Prepares and issues FRAGOs, OPORDs, OPLANs, intelligence summaries, intelligence reports (INTREPs), and situation report (SITREPs).
- Plans and orchestrates brigade briefings, debriefings and rehearsals.

Critical Item Reporting

2-104. The commander must be notified immediately of factors affecting the mission.

Friendly Factors

2-105. Friendly force factors that can affect the mission include—

- Changes in higher, subordinate, or adjacent unit mission.
- Changes in task organization.

- Changes in boundaries.
- Changes in supporting fires or CAS priority.
- Losses of unit combat effectiveness including direct support (DS) or attached units, whether maneuver, Army forces, or sustainment (maintenance and logistics).
- Critical changes in classes III and V availability or location.
- Changes in status of obstacles and contaminated areas.
- Use of smoke.
- Use of nuclear and directed-energy weapons.
- Other elements of information according to the brigade commander's guidance.

Enemy Factors

2-106. Enemy factors affecting the mission include—

- Contact with or sighting of enemy maneuver or FS forces.
- Absence of enemy forces in an area or zone.
- Movement of enemy units—withdrawal, lateral, or forward.
- Employment of the enemy's reserve.
- Use of CBRN weapons or sighting of CBRN capable equipment.
- Use of directed-energy weapons.
- Use of smoke.
- AD forces.
- Logistics stockpiles.
- Other elements of information according to the brigade commander's guidance.

TACTICAL COMMAND POST

2-107. The tactical command post is a command and control facility containing a tailored portion of a unit headquarters designed to control current operations. The TAC CP includes representatives of all the WFFs. The TAC CP is established to enhance C2 of current operations. It can be deployed to higher or subordinate headquarters facilitating parallel planning, or when distances are too extended to operate from the main CP. It must communicate with higher headquarters, adjacent units, employed subordinate units, and the main CP.

2-108. Normal TAC CP functions include the following:

- Control current operations, to include resynchronizing forces and WFFs.
- Provide information to the common operational picture.
- Monitor and assess the progress of operations.
- Monitor and assess the progress of higher and adjacent units.
- Perform targeting for current operations.
- Perform short-range planning.
- Provide input to future operations planning.
- Provide a facility for the commander to control operations, issue orders, and conduct rehearsals.
- Maintain the COP and assisting in developing situational understanding (SU).
- Analyze information for immediate intelligence.
- Develop combat intelligence of immediate interest to the commander.
- Maneuver forces.
- Control and coordinate FS.
- Coordinate with adjacent units and forward AD elements.
- Monitor and communicate sustainment requirements (classes III and V) to the main CP.

2-109. The TAC CP is small in size and electronic signature to facilitate security and rapid, frequent displacement. Its organizational layout, personnel, and equipment must be in the unit SOP.

2-110. The TAC CP is normally comprised of the command group, personnel from the S-2 and S-3, and the FSE. The S-3 section is responsible for the TAC CP. Augmentation may include—

- SP, TACOPS officer, SO, and other selected warrant officers.
- ALO, engineer, and S-9, if available.
- Representatives from the S-1 and/or S-4 (if the main CP is displacing).

2-111. METT-TC may dictate that an effective TAC CP operates from a C2-equipped UH-60.

SECTION VI – BRIGADE PLANNING

2-112. The aviation brigade is the first echelon at which the synchronization of all aspects of Army aviation operations occurs. It is comprised of staff possessing the expertise for battle command, and planning and coordination; and contains the logistics support necessary for force tailoring all types of aviation units and execution of the core competencies of aviation.

2-113. Aviation operations are inherently combined arms operations. Since aviation battalions are principally focused in a particular functional area, such as lift or attack reconnaissance, brigades are required to conduct full spectrum operations, planning, and to synchronize and orchestrate operations of different types of aviation battalions.

2-114. The aviation battalion, as the principal fighting component of the brigade, is optimized to conduct and support TACOPS. It contains the first level of staff planning, integration, coordination, and sustainment for aviation in joint or combined arms operations. It is normally the lowest level aviation unit operating independently or autonomously for any extended period of time, and then only with required support from the parent brigade, especially if task organized. The company, as the primary fighting component of the battalion, is the basic building block of aviation and is optimized for offensive actions.

2-115. The aviation brigade develops its OPLANs as an integral part of its higher headquarters staff, at its own headquarters, or both.

BRIGADE'S HIGHER HEADQUARTERS

2-116. When the aviation brigade assists the higher headquarters staff in the development of the overall plan, it saves time. The intelligence situation and ATO changes and restrictions are immediately available to all planners. Additionally, because aviation expertise is involved throughout the planning process, it ensures that aviation-related issues are resolved concurrently with plan development. All of the above preclude time-consuming queries associated with planning at different locations, thus saving critical time in developing and distributing the required orders to execute the plan.

AVIATION BRIGADE HEADQUARTERS

2-117. In addition to operational mission planning, the aviation brigade must ensure all aviation operations details are also accomplished. Those details are planned, coordinated, and rehearsed concurrently with OPLAN development. Examples of ongoing preparation include—

- Task organization actions, such as unit movements or exchange of liaison personnel.
- Airspace C2 coordination.
- TAGS, ACO, ATO, and SPINS.
- Selected rehearsals and training.
- FARP movement, composition, and emplacement.
- Maintenance support movement, composition, and emplacement.
- Weather checks and analysis.
- Passage of lines planning.
- AD status.
- Ammunition availability.
- External fuel tank distribution and management.

- Internal configuration of assault and GS aircraft.
- Communications planning.
- ASE requirements and settings.
- Identification, friend or foe (IFF) procedures and Mode 4 settings.
- PR planning.
- Brigade AA management.

2-118. Planning becomes more complicated for air assaults, insertions/extractions, and interdiction attacks on HPTs/HVTs out of contact with supported ground maneuver brigades. Operations beyond friendly lines may involve deep penetrations, wide sweeps, and bypassing enemy forces and terrain obstacles, usually at night. To react quickly to intelligence on hostile forces, planning and execution must keep pace with the accelerated attack tempo, maximizing surprise to ensure effective execution at the decisive place and time.

Aircraft Considerations

2-119. Training, planning, and operations differ between battalions due to different aircraft types organic to each. For example, although both the AHB and GSAB conduct air movement operations, missions involving heavy loads or high/hot flight environments are better left for the CH-47s in the GSAB. Refer to appendix F for additional information on aircraft characteristics.

Operational Limitations

2-120. Examples of operational limitations for helicopters and helicopter units include the following:

- Weather dictates flying some missions and deployments using instrument flight rules (IFR).
- High temperatures, humidity, altitudes, and other environmental effects reduce payloads and flight endurance.
- Weather effects (fog, heavy rain, blowing snow) or battlefield obscuration (smoke, dust) may limit day and night aided visibility and aircraft speed.
- Low ceilings limit terrain flight in mountainous or rolling terrain.
- Weather conditions (visibility, ice, high winds, and excessive turbulence) may preclude aviation operations.
- If used, auxiliary fuel tanks limit allowable ammunition loads.
- Although aerial firepower has the ability to momentarily dominate terrain, aviation units do not possess the ability to hold terrain.
- Crew endurance and aircraft maintenance requirements impact aircraft availability.
- Terrain may limit the ability to properly mask the aircraft or conduct terrain flight.

Security/Force Protection

2-121. Aviation units have limited capability to secure unit AAs while concurrently conducting operations and performing maintenance. Battlefields of a noncontiguous, asymmetric nature require aviation forces carefully consider security requirements. This kind of OE rarely has clearly defined flanks or sustainment areas. Forces must be allocated to protect critical assets against conventional and terrorist attacks. Mutual support can reduce the amount of dedicated security needed by aviation forces.

Fratricide Prevention

2-122. Six errors contributing to a fratricide incident are—

- Target misidentification.
- Inaccurate target location.
- Communication errors.
- Incorrect computations.
- Improper weapon employment.
- Mechanical malfunction.

2-123. SA, specifically timely and accurate information on friendly and enemy locations, is by far the best prevention technique. Technological advances, coordinated planning, and close communication are the best techniques to increase SA and decrease the risk of fratricide. Specific preventative measures include the following:

- Habitual relationships between ground and aviation units.
- Associated mission graphics, control measures, and ground commander's intent disseminated and understood at aircrew level.
- Distinctive and easily identifiable markings on friendly equipment using materials visible at night, such as thermal imagery tape or infrared (IR) lights.
- Fratricide prevention measures integrated into SOPs.
- FSCMs.
- Well-rehearsed plans, fully supported by tested battle drills involving all elements of the air and ground force.

2-124. Aviators may have to fly helicopters near friendly units during mission execution. Factors able to reduce potential ground and air fratricide include the following:

- Automated identification measures such as IFF, FBCB2, Blue Force Tracker (BFT), and enhanced position location reporting system (EPLRS).
- Precision-guided munitions.
- Planned or hasty coordination and control.
- Knowledge of the ground tactical plan.
- Knowledge of the exact location of friendly forces.
- Knowledge of the exact location of aircraft.
- Knowledge of friendly marking techniques.
- Positive identification of targets.
- Familiarity between the supported and aviation units.
- An AD weapon control status of weapons tight or weapons hold (FM 1-02).
  - Weapons tight-weapons systems may be fired only at targets recognized as hostile.
  - Weapons hold-weapons systems may only be fired in self-defense or in response to a formal order.

2-125. The BAE plays a key role in coordinating aviation support to the ground commander. Refer to TC 1-400 for additional BAE information. The BAE helps reduce fratricide risk by fostering communication, coordinating AC2, and keeping both aviation and ground units informed of friendly locations.

2-126. The aviation commander and division engineer coordinator ensure obstacle graphics are updated and all maneuver elements are alerted following a Volcano mission.

Logistics Support

2-127. The combination of the OE and the diversity of the aviation brigade's battalions often require FARPs and maintenance teams to operate simultaneously at different locations. Establishment and resupply operations require careful planning and coordination. When possible, these activities should be part of the mission rehearsal.

Mission, Enemy, Terrain and Weather, Time Available, Troops and Support Available and Civil Considerations

2-128. Planning considerations are predicated on METT-TC. Some of these elements are specific to the mission and are discussed in chapter 3. This section addresses planning considerations common to any mission the brigade might be assigned.

Mission

2-129. Higher headquarters assign missions to the aviation brigade. Commanders determine their specified and implied tasks by analyzing their assigned mission and coordinating with supported units. Results of this analysis yield essential tasks that, together with the purpose of the operation, clearly indicate the actions required. The mission includes which tasks must be accomplished; who is to complete them; and when, where, and why they are to be conducted. It also includes CRM considerations.

Mission Criteria

2-130. For any mission, the commander seeks to establish criteria maximizing his or her probability of success (ground conditions, visibility, and force ratios). The supported commander and brigade higher headquarters set mission criteria. During the planning process, mission criteria are quantified and stated in easily understood terms. If any of the stated criteria are achieved before or during the mission, the designated commander should execute predetermined actions. Following are several considerations influencing mission criteria.

- **Weather**. AR 95-1 sets minimum weather conditions, stated as ceiling and visibility, for certain types of helicopter missions over certain types of terrain. Weather conditions must be at or above minimums for the entire time aircraft are flying and over the entire area in which they are operating, unless waived by the Commanding General due to criticality of a specific combat operation. Commanders may establish minimum weather requirements above those stated in AR 95-1.
- **Aircraft available**. Mission effectiveness with minimum casualties requires rapid massing of combat power at the critical place and time. If the mission requires more aircraft than available (either due to combat loss, nonmission-capable aircraft, or lack of aircrews), then the commander may terminate the mission.
- **Time**. If mission delays mean aircraft cannot apply required combat power at a specified hard time, the commander may modify or terminate the mission.
- **Lack of mission-essential combat power**. Possibly the result of increased enemy or decreased friendly capability. If the attack reconnaissance force meets stronger than expected resistance or loses combat or supporting assets en route, the commander may request additional division, corps, or joint support before modifying or terminating the mission.
- **Mission criticality**. The importance of the mission drives the abort criteria. Less critical missions are quicker to terminate. For example, attack reconnaissance aircraft may perform corps or higher echelon shaping operations. While some operations may depend on the success of an attack reconnaissance mission, others may be harassing in nature and not as critical to the campaign.
- **Enemy**. Enemy activity along flight routes or in battle positions (BPs) resulting in extensive friendly losses may require a mission termination. Critical joint mission needs that divert supporting fires may arise just before or during a mission. A catastrophic event, such as a nuclear explosion or unexpected use of chemical weapons, may also cause higher headquarters or the commander to terminate or modify the mission.

Mission Modifications

2-131. The battalion assists in planning actions and reactions for these situations, but specific modification or mission termination criteria must be set prior to execution. If any of the stated criteria are achieved before or during the mission, the air mission commander (AMC) must be prepared to advise the commander. Example actions include delay, divert, or terminate in part.

- **Delay**. If sufficient time remains and circumstances can reverse with ground combat and other supporting fires, the commander may delay a mission. He or she may place aircraft in a "racetrack" pattern, reduce their airspeed, or land them in holding areas (HAs). The commander might decide to take similar actions if forecast weather suddenly changes forcing aircraft to land or proceed at slower airspeeds and/or lower altitudes.

- **Divert.** If time, fuel, or safe laager areas are not available to permit a delay, the commander may execute a divert contingency. Examples include use of alternate flight routes to avoid threats or foggy areas and use of alternate BPs.
- **Postpone or terminate in part.** If a situation exists that a delay or divert cannot correct, the commander may decide to postpone or terminate a mission phase and attempt to continue with available forces and support.

Enemy

2-132. Analysis of the enemy includes information about its strength, location, activity, and capabilities. Commanders and staffs also assess the most likely enemy COAs. Analysis includes adversaries, potentially hostile parties, and other threats to success. Threats may include the spread of infectious disease, regional instabilities, or misinformation. Commanders consider asymmetric as well as conventional threats.

2-133. Brigades conduct a threat analysis during planning based upon the IPB prepared with higher headquarters. A common mistake is orienting too much on terrain as opposed to the enemy. Knowing the enemy's location, its forces, capabilities, and intentions are key to success. Knowledge of the enemy ensures the best use of terrain to exploit its weaknesses and capitalize on friendly strengths.

2-134. ASE settings depend on accurately analyzing the enemy AD threat. Knowing the threat is critical to effective passive and active countermeasures.

Terrain and Weather

2-135. Terrain includes manmade features such as cities, airfields, bridges, railroads, ports, and contaminated areas. Terrain and weather also have pronounced effects on ground and air maneuver, precision munitions, air support, and sustainment operations. To find tactical advantages, commanders and staffs analyze and compare environmental limitations on friendly, enemy, and neutral forces.

Terrain Analysis

2-136. Commanders and staffs perform terrain analysis whether using digitized tools or paper maps. They evaluate terrain for cover and concealment, its impact on maneuver, and the enemy's movements. The key elements of terrain analysis are summarized in the following mnemonic OAKOC:

- Observation and fields of fire.
- Avenues of approach.
- Key terrain.
- Obstacles.
- Cover and concealment.

Obstacles

2-137. Obstacles and reinforcement of terrain must be included in the tactical plan. Engineers use obstacles to disrupt, fix, turn, or block the enemy. Disruptive obstacles cause enemy formations to separate or bunch up, which disrupts their maneuver and attack. Fixing obstacles slow enemy progress and allow friendly fires the opportunity to mass effects. Turning obstacles drive the enemy toward friendly engagement areas (EAs) and massed fires, or force the enemy to expose its flanks. Blocking obstacles deny the enemy access to an area or prevent advance in a given direction. Although the brigade probably will not have engineer support to establish obstacles, the commander must understand the ground force commander's obstacle plan and use it to his or her advantage.

Terrain Reconnaissance

2-138. Because maps are sometimes inaccurate or incomplete, commanders should conduct detailed, personal reconnaissance. Brigade commanders should create conditions where battalion commanders can ensure their aircrews are familiar with the terrain and scheme of maneuver. If possible, battalion commanders—and their crews—should perform a map reconnaissance; visit landing zones (LZs), PZs, BPs, and firing positions (FPs); and conduct rehearsals. These actions help crew members understand the scheme of maneuver and commander's intent, and quicken their reactions during the chaos of battle. Commanders consider all sources of intelligence. Aerial photographs, satellite imagery, and human intelligence can be critical.

Weather

2-139. Weather affects Soldiers, equipment, operations, and terrain. Cloud cover, wind, rain, snow, fog, dust, light conditions, and temperature extremes combine in various ways to affect human efficiency. They also limit the use of weapons and equipment. Weather impacts both friendly and enemy assets. For example, rain can degrade thermal imaging systems (TISs), but it also inhibits the cross-country maneuverability of enemy forces. Each system used on the battlefield has its strong and weak points in relation to the weather. Commanders must know the strengths of their systems and use them to attack the weaknesses of enemy systems.

Visibility

2-140. Limited visibility affects operations and often favors ground maneuver. Fog and smoke reduce the effective range of many weapon systems, including AD weapons. Commanders use concealment of limited visibility to maneuver forces to a positional advantage. The brigade should plan operations maximizing advantages of its superior sensor systems.

Troops and Support Available

2-141. Commanders assess the training level and psychological state of friendly forces. This analysis includes availability of critical systems and joint support. They examine maneuver, Army forces, and sustainment assets including contractors. The status of all aviation brigade units should be readily available for the commander and staff in accordance with SOP.

Supporting Fires

2-142. The brigade will frequently have access to supporting fires from a coordinated fires network. These complementary fires could facilitate movement to the objective through J-SEAD, engage targets bypassed by aircraft, or provide indirect fires on the objective. Knowing what type of FS is available, and when it is available, are important considerations during mission planning and EA development. Efforts to coordinate joint fires for actions on the objective could be critical toward success of long-range operations.

Airspace Coordination

2-143. Total familiarity with the TAGS is essential to deconflict operations and prevent mission delays. Brigades may need to comply with provisions in the ACO, ATO, and SPINS. They have strict timelines and FSCMs to take into account during brigade and subordinate planning cycles.

2-144. An ACO is an order implementing the airspace control plan by providing details of the approved requests for ACMs. It is published either as part of the ATO or as a separate document. ACO coordination is required operations outside Army controlled airspace.

2-145. An ATO is a method of tasking and disseminating to components, subordinate units, and C2 agencies projected sorties, capabilities, and/or forces to targets and specific missions. It normally provides specific instructions including call signs, targets, and controlling agencies as well as general instructions.

Air Ground Integration and Coordination

2-146. The BAE is an aviation planning and coordination cell organic to the BCT that synchronizes aviation operations into the ground commander's scheme of maneuver. Working in conjunction with the BAE, the aviation brigade must ensure all aspects of the mission are thoroughly planned, coordinated, and rehearsed with the supported unit. Supported unit graphics are essential for SU. Aviation often conducts passage of lines with supported units, and those operations require close coordination. Fires are considered to ensure necessary artillery is available when called. When appropriate, aviation brigade liaison teams augment the BAE in coordinating and executing aviation missions for the BCT.

Time Available

2-147. Commanders assess time available for planning, preparing, and executing the mission. They consider how friendly and enemy forces will use the time and the possible results. Proper use of time available can be a key to success. The one-third/two-third rule is used whenever possible. Concurrent planning makes the best use of time. Emerging digital systems enhance concurrent planning capabilities.

Civil Considerations

2-148. Army forces operate among populations with diverse cultures and political orientations that may support, oppose, or remain ambivalent to their presence. Dealing with local populations requires assessing a myriad of factors. These are civil considerations. Although some civil considerations are mechanistic and predictable (such as electric power grids and railway networks) most include the human dimension. Economies, political systems, and social institutions exhibit willful behavior that is unpredictable and unruly. Understanding how civil considerations interrelate facilitates understanding direct and indirect consequences of actions. For this reason, IPB includes an analysis and evaluation of civil considerations.

2-149. Civil considerations involve the impact of the local populace on operations and the impact of operations on the local populace. At higher levels, civil considerations also include larger, long-term diplomatic, informational, and economic issues. Civil considerations are comprised of six characteristics—areas, structures, capabilities, organizations, people, and events (expressed in the memory aid ASCOPE). Refer to FM 3-0 and FM 6-0.

Civil Impact

2-150. Civil considerations at the tactical level generally focus on the immediate impact of civilians on current operations; however, they also consider larger, long-term diplomatic, economic, and information issues. Civil considerations can tax the resources of tactical commanders. The local population and displaced persons influence commanders' decisions. Their presence and the need to address their control, protection, and welfare affect the choice of COAs and allocation of resources. In stability operations, civilians can be a central feature of planning.

Political Boundaries

2-151. Political boundaries of nations, provinces, and towns are important considerations. Conflict often develops across boundaries, and boundaries may impose limits on friendly action. Boundaries, whether official or not, determine which civilian leaders and institutions can influence a situation.

Media Presence

2-152. Media presence guarantees a global audience views military activity in near real-time. The activities of the force—including individual Soldiers—can have far-reaching effects on domestic and international opinion.

PLANNING MODELS

2-153. Aviation brigades plan missions to support ground units. An air assault is an example of a mission in support of a ground unit. Aviation brigades also plan missions commanded and controlled by the aviation brigade. An attack by aviation forces across the forward line of own troops (FLOT) using attack reconnaissance helicopters is an example of a mission under the C2 of the aviation brigade.

Common Planning Process

2-154. The planning process for aviation brigade operations does not differ from the doctrinal processes already in place. Because the brigade may have units joining it from each aviation mission area, it is critical to discuss the common factors and differences each brings to the brigade. Critical planning includes reconnaissance, security, CCA, interdiction attacks, air assault, air movement, aerial mine emplacement, AD, AC2, FS, C2, ATS, FARP operations, and aeromedical evacuation. Brigade planners may be available from each aviation mission area. If not available, planners must still plan missions to the same level of expertise and detail expected of a mission area SME.

Reverse Planning Process

2-155. Planning begins with the terminal end of the mission—actions at the objective, the cargo delivery point, and the passenger drop-off point. Table 2-1, page 2-26, shows the commonality of the planning phases of each mission area. It is intended as a starting point to assist in team building.

Table 2-1. Planning phases

Air Assault	*Attack*	*Recon and Security*	*Air Movement*	*C2*
Ground tactical plan.	EA plan.	Observation and engagement plan.	Pax and cargo delivery plan.	C2 support plan.
Landing plan.	BP/HA occupation plan.	Recon plan.	Landing plan.	Landing plan.
Air movement plan	Movement techniques	Movement techniques	Movement techniques	Movement techniques
Loading plan (Pax & equip).	Loading plan (ammo).	Loading plan (ammo).	Loading plan (Pax & cargo).	Loading plan (Cdrs & staff).
Staging plan (pickup zone [PZ]).	Staging plan (forward assembly area [FAA]).	Staging plan (FAA).	Staging plan (PZ).	Staging plan (PZ).

PLANNING RESPONSIBILITIES

2-156. For most operations aviation brigade and subordinate battalions plan at different levels. Table 2-2 provides a general guide for planning responsibilities.

2-157.

Table 2-2. Brigade and battalion planning responsibilities

Aviation Brigade provides Battalion	*Aviation Battalion determines*
General timings.	Exact speeds, routes, flight modes, & timings.
H-hour (line of departure [LD], LZ).	Adjustments as LD time nears.
Passage point (PP) locations.	Exact flight route.
Suppression of enemy air defense (SEAD) /J-SEAD plan.	Exact planning times from AA to LD, PP, BP, PZ, or LZ.
EAs, LZs, PZs, areas of potential BPs.	Exact flight routes.
Flight axes.	Exact surveillance plan
Named areas of interest (NAIs)/target areas of interest/decisive points (DPs).	Release points (RPs), rally points, FPs, attack by fire positions, exact BPs, kill zones, landing areas

2-158. Figures 2-2, page 2-27, and figure 2-3, page 2-28, graphically depict the planning responsibilities for air ground integration and incorporate the general rules in table 2-2. They also include some of the planning steps of the aviation brigade's higher headquarters.

2-159. Figure 2-4, page 2-29, depicts an attack by the aviation brigade forward of the FLOT. Figure 2-5, page 2-30, depicts an air assault supported by the aviation brigade. Figure 2-6, page 2-31, depicts an aviation brigade supporting a ground brigade both in front of and behind the FLOT.

2-160. Times and airspeeds depicted in these figures are examples. Additionally, circumstances may require the brigade provide the exact routes (airspace coordination) and times to affect timely coordination with supporting elements.

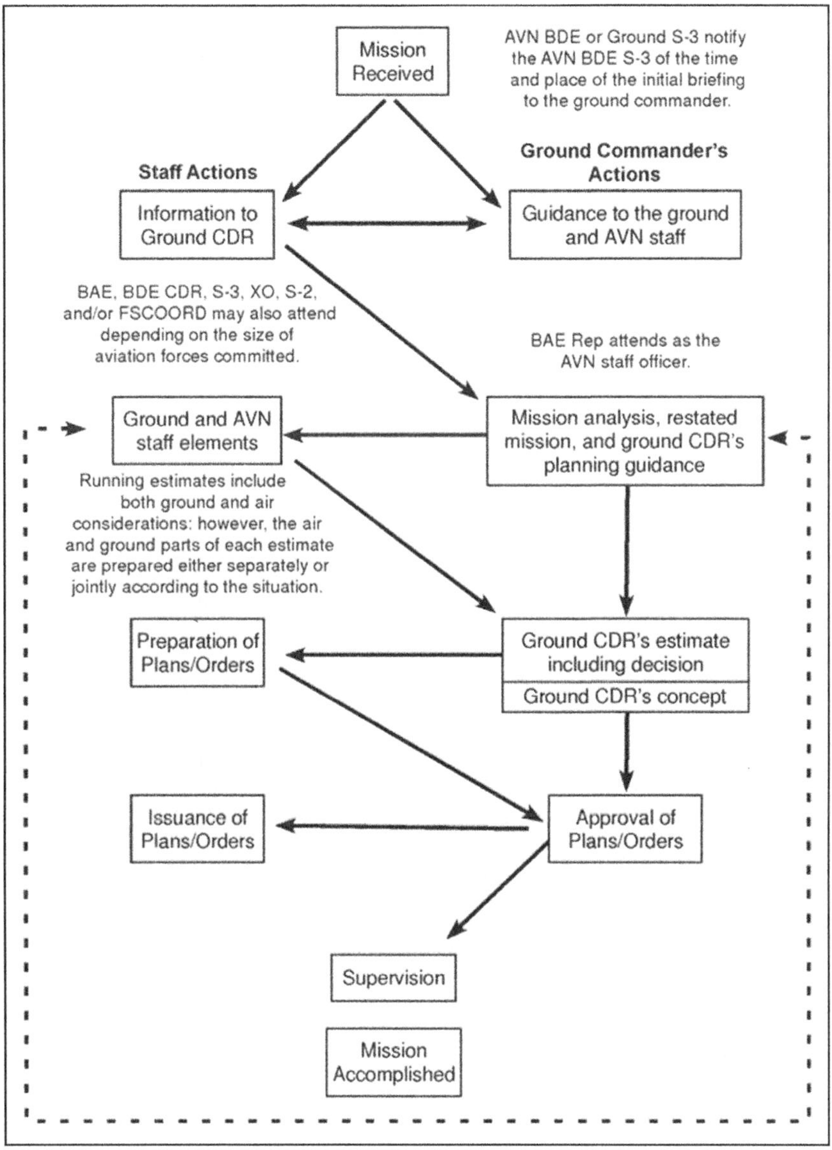

Figure 2-2. Brigade planning responsibilities, aviation forces in support of a ground unit

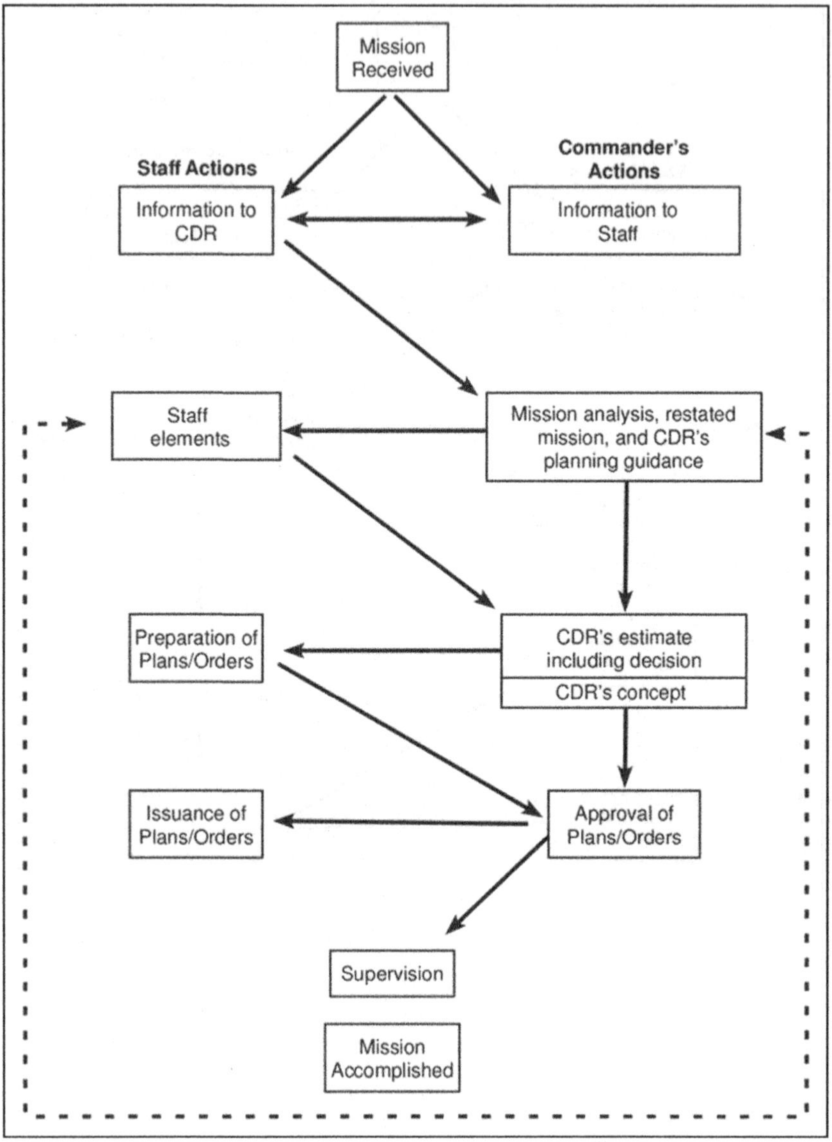

Figure 2-3. Brigade planning responsibilities, aviation forces under aviation brigade control

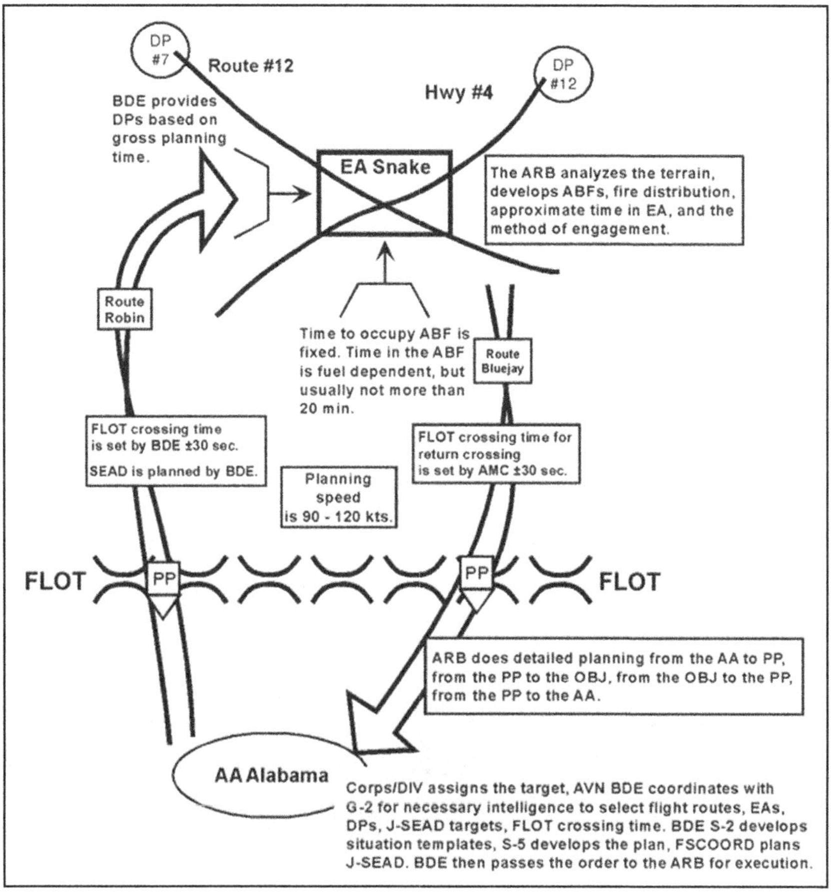

Figure 2-4. Aviation brigade conducts an attack

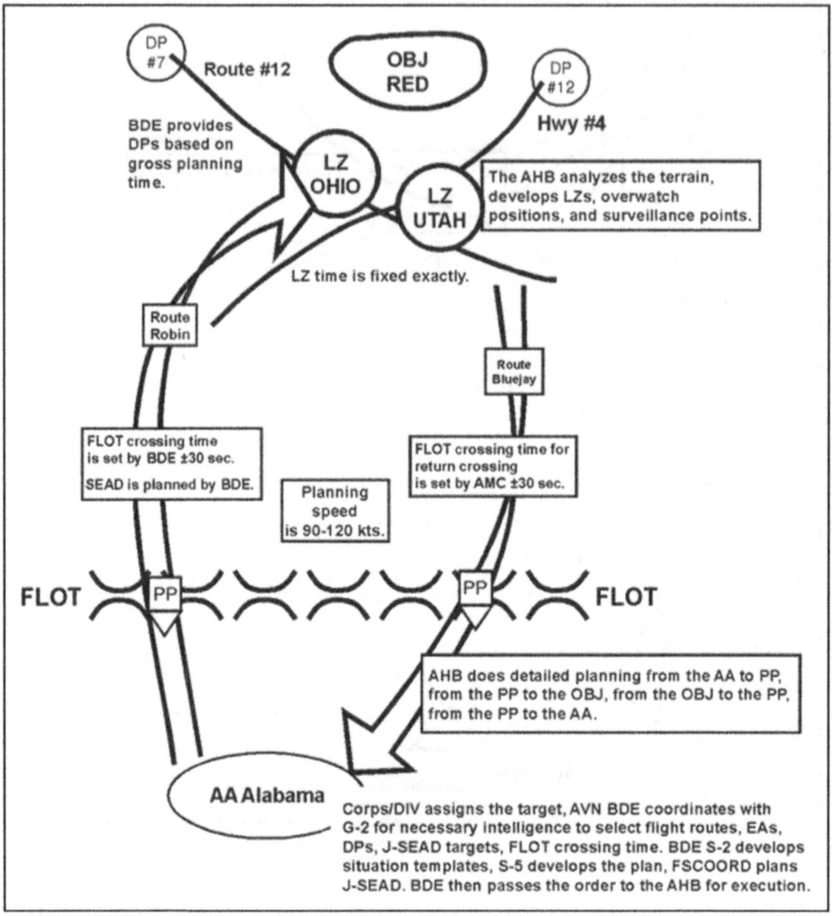

Figure 2-5. Aviation brigade conducts an air assault

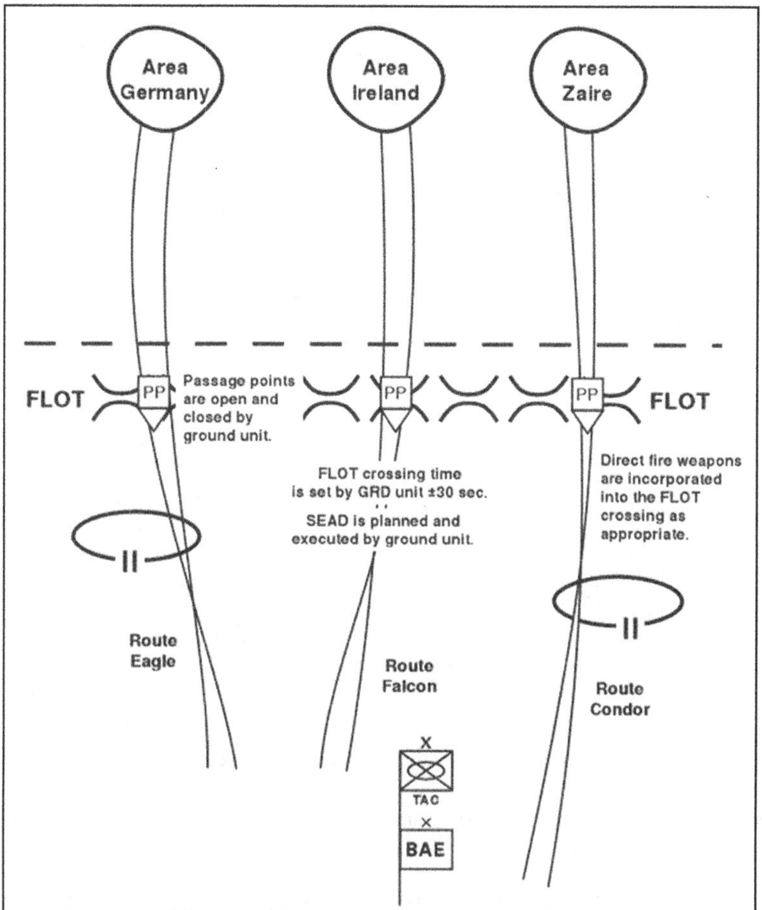

Area Germany All operations within the ground BDE sector are under the control of the ground unit CDR; however, aviation operations within his sector are planned through a cooperative effort by the AVN BDE and ground unit staffs facilitated by the BAE. The end result is a ground unit order that is written by the ground unit with AVN BDE input.

Area Ireland The staff planning process for combined air and ground operations in a ground unit sector follows the same process as a ground operation except that the AVN BDE staff and BAE assist in the planning and preparation of the order.

Figure 2-6. Aviation brigade supports ground brigade operations

MILITARY DECISIONMAKING PROCESS

2-161. To plan and coordinate missions effectively, the commander and staff follow the MDMP (figure 2-7). Staff planners must focus on previously listed aviation planning considerations to formulate a complete plan. Because of the complexity inherent in the process, regular MDMP exercises are essential prior to deployment. FM 5-0 and FMI 5-0.1 discuss this process in detail.

2-162. The dynamic battlefield does not often allow a complete MDMP because of time constraints. The commander and staff must know current operational readiness in order to assess feasibility of mission requests immediately.

2-163. The steps of an abbreviated MDMP are the same as those for the full process; however, the commander performs many of them mentally or with less staff involvement (figure 2-7). The commander may direct a COA based on experience to expedite planning. The products developed during an abbreviated MDMP may be the same as those developed using the full process; however, they are usually less detailed. Some may be omitted altogether. Unit SOPs should address how to abbreviate the MDMP based on the commander's preferences.

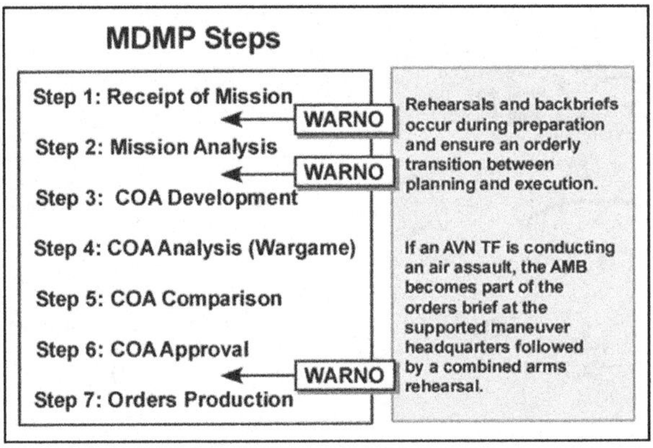

Figure 2-7. Military decisionmaking process

OPERATIONS IN CONTIGUOUS AND UNASSIGNED AREAS

2-164. Mission planning factors vary between operations in contiguous and unassigned areas. These differences include the following:

- Graphics for contiguous operations normally include restrictive control measures to prevent fratricide during engagements in close proximity to friendly elements.
- Actions on the objective for the two operations differ. In a close fight, the attack reconnaissance unit maneuvers in concert with ground elements to observe and deliver ordnance on the enemy force. With an enemy out of contact with ground forces, aviation units conduct a movement to contact to develop the situation.
- Weapons loads for AH-64s may vary as auxiliary fuel tanks are added for extended range operations. Further, commanders may choose to include auxiliary fuel tanks on aircraft in close fights increasing station time at the objective area.
- Operations in unassigned areas involve coordinated planning with higher echelons and joint elements, while operations in close proximity can be organized at division and BCT levels. For instance, unassigned area operations primarily involve joint fires while contiguous operations use ground unit direct fires, FA support, and joint fires.

- C2 headquarters for close fights are normally located with the ground element in contact. Operations out of contact with ground units are normally controlled from the aviation brigade.

Aviation Mission Planning System

2-165. Aviation mission planning system (AMPS) is an automated mission planning and synchronization tool designed specifically for aviation operations. Aviation brigade, battalion, and company flight planning sections or CPs use AMPS to automate all mission planning tasks, including functions such as tactical planning, mission management, and aircraft data loading. Portions of AMPS may be used as part of the rehearsal process.

Tactical Planning Function

2-166. The tactical planning function of AMPS includes brigade- and battalion-level planning tasks, such as intelligence data processing and route, communications, and navigation planning. This facilitates review and preparation of the air mission brief. Additional AMPS uses are—

- Creation of detailed terrain analysis.
- Determination of line of sight (LOS) and intervisibility between a BP and an EA.
- Determination of prominent terrain along the route to be flown, using the perspective view feature.
- Creation and distribution of graphics.

2-167. The BAE and division CPs have AMPS available to assist in COA development and wargaming during the MDMP, reverse-planning, and coordination processes. During air assaults, the ground maneuver air assault task force (AATF) staff can exploit AMPS to simplify preparation of the landing, air movement, and loading plans. The brigade may employ AMPS to plan shaping operations and integrate aviation routes with other J-SEAD and shaping fires.

2-168. Because BAE, aviation brigade, and battalion and below planners have AMPS access, planning can occur concurrently. Planners can use AMPS to pass aviation brigade and ground maneuver planning to lower echelons updating their plans. The orders function of AMPS assists OPORD, WARNO, and FRAGO development and distribution to lower echelons. This facilitates passing up-to-date information and changes from higher headquarters and supported units.

2-169. The mission management function also facilitates company and platoon level planning. These tasks include aircraft performance planning, weight and balance calculations, flight planning, and fighter management. They also include OPLAN changes and OPORD development. It helps companies and platoons conduct rehearsals using the route visualization and intervisibility features of AMPS.

SECTION VII – AVIATION BRIGADE COMMUNICATIONS

2-170. Combat information reporting and its exploitation are fundamental to combat operations. This information and the opportunities it presents are of interest to other maneuver units and higher headquarters staffs. Combat information reporting requires wide and rapid dissemination. Brigade elements frequently operate over long distances, wide fronts, and extended depths from their controlling headquarters. Communications must be redundant and long range to meet internal and external requirements. Long-range communications can be augmented through signal support. The systems must be in place before they are needed.

2-171. Communication is a major challenge for the aviation brigade. Operations in close proximity to the enemy require terrain flight altitudes that make LOS communications difficult. CPs and aircrews may employ radio relay, retransmission, or alternate communications to maintain contact. High frequency (HF) radio provides alternate non-line of sight (NLOS) communications for longer distance missions and nap-of-the-earth (NOE) communications. Satellite communications (SATCOM) is available to support both C2 aircraft customers and the brigade's own C2 needs. When UAS have the capability of conducting radio relay and retransmission missions they may enhance and ensure better communications over extended distances.

HIGHER TO SUBORDINATE

2-172. Brigade headquarters ensures its communications architecture (command, operations and intelligence [O&I], administrative and logistics [A&L], FS, and SATCOM) is operational at all times. The retransmission system is dedicated to on-call restoration of communications on any net. Possible retransmission locations must be identified and checked before starting operations. The brigade has signal support from higher headquarters during operations.

SUBORDINATE TO HIGHER

2-173. Battalions and separate companies continually monitor brigade nets as directed (usually command and O&I). Likewise, the brigade continually monitors its higher headquarters nets.

SUPPORTING TO SUPPORTED

2-174. Liaison elements supporting the brigade maintain communications between their organization and the aviation brigade.

COMMUNICATIONS DISRUPTION

2-175. Communications, particularly electromagnetic, are subject to disruption. This may result from unintentional friendly interference, intentional enemy action, equipment failure, atmospheric conditions, electromagnetic pulse, or terrain interference. To compensate for these, the commander should—

- Provide for redundancy in means of communication.
- Ensure subordinates understand commander's intent.
- Avoid overloading the communications systems.
- Minimize use of radio.
- Ensure signal security and COMSEC practices are followed.

COMMUNICATION RESPONSIBILITIES

2-176. All levels of command gain and maintain communications with necessary headquarters and personnel. Communications methods and procedures should be established in unit SOPs and practiced during battle drills and flight operations. Traditional communications responsibilities are—

- **Higher to lower**. The higher unit establishes and maintains communications with a lower unit. An attached unit of any size is considered lower to the command to which it is attached.
- **Supporting to supported**. A supporting unit establishes and maintains communications with the supported unit.
- **Reinforcing to reinforced**. A reinforcing unit establishes and maintains communications with the reinforced unit.
- **Passage of lines**. During passage of lines (forward, rearward, or lateral), the passing unit establishes initial contact with the stationary unit. However, the primary flow of information must be from the unit in contact.
- **Lateral communications**. Establishing communications between adjacent units may be fixed by the next higher commander, by order, or by SOP. If responsibility is not fixed, the commander of the unit on the left establishes communications with the unit on the right.
- **Rear to front communications**. The commander of a unit positioned behind another unit establishes communications with the forward unit.

2-177. Regardless of establishment responsibility, all units take prompt action to restore lost communications.

AIRBORNE COMMAND AND CONTROL

2-178. Inherent in the brigade mission is transport of commanders and staff officers allowing them to see the battlefield and more effectively C2 their units. A2C2S equipped aircraft, found in the GSAB, can serve as an airborne TAC CP with the same digital capabilities as the ground TAC CP. Appendices B and E provide additional information.

Aerial Reconnaissance

2-179. Brigade elements may be employed to verify enemy and friendly unit locations or even their existence. For example, if the higher headquarters commander loses communications with a subordinate unit, that commander may ask the aviation commander to verify the unit's location and status.

Video Teleconference

2-180. Video teleconferences (VTCs) among corps, divisions, and brigades are becoming more common. Some brigades are already fielded with this capability. VTCs save commanders time and are an excellent method of long distance coordination.

Wire

2-181. Normally wire is used for communications within the CP, AA, and support area. It is the primary and most secure means of communication whenever the situation permits. Initially, wire is laid on the ground. Then, if time permits, wire is buried or installed overhead. Buried wire is the preferred method to counter enemy intrusion and electromagnetic pulse. However, wire should be overhead when crossing roads, except where culverts and bridges are available. Overhead wire should be a minimum of 18 feet (5.5 meters) above ground. Wire should be tagged according to a system in the SOP. At a minimum, tags should be at the ends of each line. This facilitates reattaching wires when they are pulled out or cut. Use of overhead wire in the vicinity of helipads and airfields should be avoided; however, if used, they must be clearly marked.

Telephone

2-182. Telephone, cellular telephone, and satellite telephone are convenient means to communicate unclassified information. Commercial lines are used when approved by higher headquarters. It is inadvisable to rely on these means heavily during training due to limited security and availability on the battlefield. To deny enemy collection efforts, secure devices should be used with commercial lines. If a unit is forced to withdraw, and with the approval of higher headquarters, existing wire lines (including commercial lines) are cut and sections removed to prevent use by the enemy. Commanders must ensure their Soldiers do not transmit sensitive information such as Soldier mishaps, unit locations, or unit strength over cellular telephones or other non-secure means of communication.

Computers

2-183. Computers allow the exchange of intelligence, intent, orders, plans, and direction in a timely manner. The mission and structure of the brigade determine specific information flow and processing requirements. In turn, the brigade's information requirements dictate the general architecture and specific configuration of the communications and computer systems. Unit SOPs should address the use of computers.

Radio

2-184. Operations often depend on radio as the primary means of communication. This is especially true during mobile combat operations. Radio communications should be kept to an absolute minimum until enemy contact is made.

2-185. Frequency modulated (FM) communications are the primary O&I and A&L nets, and the means of communicating with ground forces. However, aviation has a broad range of other radios facilitating joint, internal, long-range, and NOE communications. Appendix B discusses the following systems:

- High frequency (HF) for long distance and NOE communications.
- Ultra high frequency (UHF) for internal communications and communication with joint aircraft.
- Very high frequency (VHF) for internal communications and communications with ATS.
- Tactical satellite and SATCOM for long distance communications.

2-186. To avoid detection by enemy direction-finding equipment, the brigade uses all other means of communication to supplement radio. Although secure equipment may prevent the enemy from knowing the content of communications, location and volume are easy to detect and analyze giving the enemy valuable combat information.

Radio Retransmission/Relay

2-187. Brigade retransmission stations are employed according to the tactical situation providing FM radio communications between stations too far apart to communicate directly. Brigade can deploy both ground and air retransmission stations. Ground retransmission normally supports the brigade command net. Airborne retransmission has a limited time on station, but is a vulnerable asset. Preplanning is essential to the effective use of airborne retransmission. Moving ground retransmission by sling-load is an efficient and effective method of emplacing radio retransmission.

2-188. The brigade can insert and resupply ground retransmission teams into sites inaccessible by ground. Brigade aircraft may carry retransmission equipment, relay equipment, or both. Aircrews can transmit or relay with onboard equipment.

Messengers

2-189. Messengers may be used anywhere, but normally are used for critical communications between CPs and higher and lower headquarters. Messengers are used during electronic and radio silence as well. While ground messengers are slower than other means of communications, aviation provides a rapid capability if preplanned. Aviation messengers may be particularly useful in carrying A&L messages when en route to and from rear units. They can be used even if units are in contact and especially when jamming or interception hampers radio communication. During electronic and radio silence, opening and closing flight plans by land lines may be required to control helicopter movements.

Message and Document Delivery

2-190. Electronic transmission of messages and documents may not be possible due to nuclear weapons employment, enemy jamming operations, imposition of radio silence, or inoperable equipment. Messages and documents warranting aerial delivery include combat plans and orders; written coordination and control measures; and graphics. They also include public affairs materials to sustain public understanding and support for the Army's continued operations. Using aviation to deliver messages or documents is a sound technique; however, it is most efficient when there is a prepared plan for execution. If an aviation messenger service is anticipated, it should be part of the aviation brigade and higher headquarters SOPs.

Tactical Internet

2-191. When digitized systems are fielded, units may receive missions via the tactical internet (TI) along with OPORDs and supporting information (appendix B). This accelerates the planning cycle and allows swift interaction among supporting elements.

Visual and Audio

2-192. Visual and audio signals are found in the SOI or SOP. The SOP may establish signals not included in the SOI. Commanders and staff planners carefully determine how sound and visual signals will be used and authenticated. Sound and visual signals include pyrotechnics, hand-and-arm, flag, metal-on-metal, rifle shot, whistles, horns, and bells. Visual cues are especially valuable in the FARP.

Intelligence, Surveillance, and Reconnaissance

2-193. ISR is distinct from the larger framework of information support as it focuses primarily on the enemy. Poor intelligence has been the immediate cause for innumerable defeats. Inadequate surveillance and reconnaissance are prime contributors. Conversely, excellent intelligence breeds bold action that negates enemy superiority. Normally, timely and accurate intelligence depends on persistent surveillance and aggressive, efficient reconnaissance.

2-194. The brigade is a key supplier of ISR; however, it is also a consumer of BCT and higher echelon (Army, joint force, and national) ISR products. By its tie-in to higher echelon ISR information, brigade executes its mission in an environment characterized much more by what is known rather than unknown.

COMMUNICATION NETS

2-195. Each aviation brigade communicates by one or more of the following systems:

- LAN (secure and nonsecure).
- Amplitude modulated (AM)/FM radio.
- HF radio.
- SATCOM.
- Maneuver control system (MCS)/FBCB2.
- Commercial lines.
- Wire.

RADIO NETS

2-196. Brigades normally operate on their own and their higher headquarters command, O&I, and A&L nets. Aviation brigades also operate on fire nets. Additionally, each aviation brigade must often monitor lower, adjacent, and supported unit radio nets. This can be especially valuable when supporting and conducting air assaults and close fires.

2-197. Critical higher headquarters radio nets must be monitored at all times.

- Higher command net. The brigade commander, all brigade CPs, and the S-3 enter and operate.
- Higher O&I net. The S-2 and all brigade CPs enter and operate.
- Higher A&L net. The S-1 and S-4 enter and operate.
- Other staff sections and officers enter other nets as appropriate.

Brigade Command Net

2-198. A secure command net, controlled by the S-3, is used for C2. All subordinate and supporting maneuver and sustainment units normally operate in this net. As a rule, only commanders, XOs, or S-3s communicate on the net (figure 2-8, page 2-38).

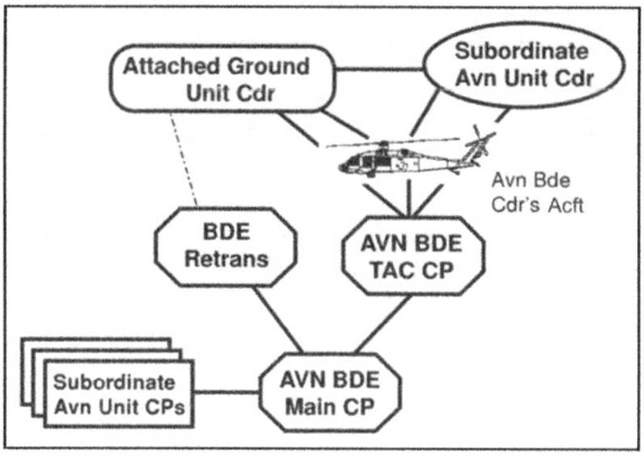

Figure 2-8. Brigade command net

Brigade Operations and Intelligence Net

2-199. The O&I net is controlled by the S-2 and is used for routine operations and INTREPs). The O&I net is not normally monitored by all elements of the brigade or subordinate staff. This net provides details and discussion that lead to analysis. When completed, this analysis is relayed to the appropriate commander. The S-2 and XO, operating in the main CP, ensure this analysis is conducted and relayed in a timely manner by appropriate means (figure 2-9). It also functions as a surveillance net when required.

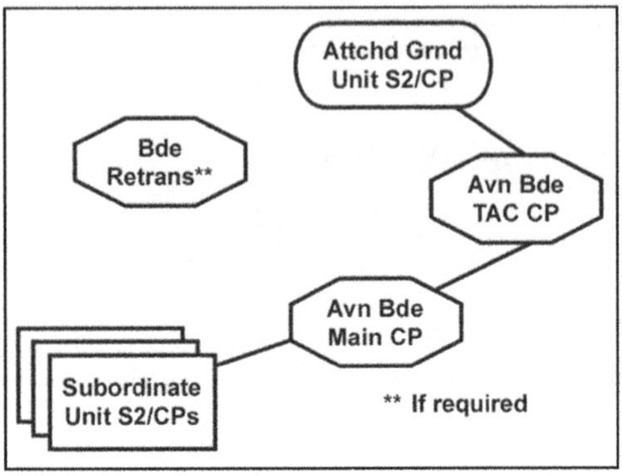

Figure 2-9. Brigade operations and intelligence net

Brigade Administrative and Logistics Net

2-200. This net is controlled by the S-1 and S-4, and is used for A&L traffic. The A&L net, like the O&I net, is normally not monitored by all elements of the brigade or subordinate commanders. The net is used for details and discussion leading to the resolution of administration and logistics matters. Critical information is relayed to the appropriate commander or discussed on the command net. The XO, operating in the main CP, ensures analysis is conducted and relayed in a timely manner by appropriate means (figure 2-10, page 2-39).

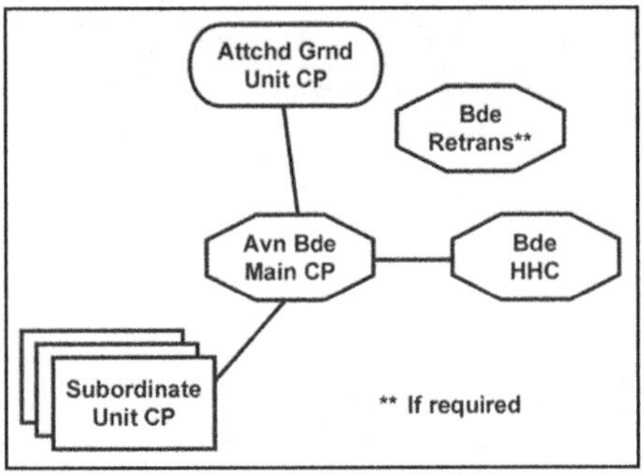

Figure 2-10. Brigade administrative and logistics net

Combat Aviation Net

2-201. The AMC, AATF commander , infantry force commander, and pickup zone control officer (PZCO) use this secure FM net for air-to-ground communication at the PZ/LZ, and to transmit SITREPs and mission changes. Aviation forces monitor this net during air assault operations and selected air movement operations, especially in the vicinity of the PZ/LZ.

Fire Control Nets

2-202. The FSCOORD operates in the supporting FA command net and a designated fire direction net to coordinate artillery fires. USAF ALO, when attached, CAS through a USAF CAS request net (HF/single side band) and a UHF/AM air-ground net.

Monitored Radio Nets

2-203. Aviation brigades must monitor the nets of subordinate, adjacent, supporting, or supported units often. This can be especially valuable in complex or fast moving operations.

2-204. The brigade commander should have three FM nets, one UHF/VHF net, one HF radio, and one SATCOM radio. These nets enable him or her to monitor subordinate unit, supported unit command, O&I, A&L, or other nets deemed important to the mission at hand.

Digital Situational Awareness Net

2-205. The digital SA net is designated primarily for constant transmission of digital information on the battlefield. Digital information traffic shares the net with digital communications. The brigade AMPS is the net hub and each battalion links to the brigade AMPS. The brigade AMPS collects spot reports (SPOTREPs), SITREPs, status reports (STATREPs), and BDA reports.

Standard Army Management Information Systems Nets

2-206. Standard Army Management Information Systems (STAMISs) consist of computer hardware and software systems automating diverse functions based on validated customer requirements and facilitating the vertical flow of logistics and maintenance status information to units Army wide. Chapter 5 addresses STAMIS architecture.

Army Battle Command System Nets

2-207. Appendix B addresses digitized unit nets.

SECTION VIII – COMMAND POST SURVIVABILITY

2-208. CPs present electronic, thermal, acoustic, visual, and moving-target signatures easily detected. Upon detection, CPs can be destroyed through overt enemy action or disrupted and exploited by electronic means unless measures are taken to reduce vulnerability. Measures include—

- Maintaining local security.
- Locating on reverse slopes to deny enemy direct and indirect fire effects.
- Locating in urban areas to harden and reduce IR or visual signatures. Collateral damage to the local population must be considered if exercising this option.
- Remotely locating and dispersing antennas.
- Dispersing CP sub-elements.
- Displacing as required by METT-TC.
- Using low probability of interception techniques—landlines, directional antennas, and messengers.
- Providing COMSEC.

2-209. In most cases, survivability requires commanders to employ a combination of the protective measures listed above. Commanders must balance employment of these measures against maintaining CP operations and effectiveness. Frequent displacement, for example, might reduce the vulnerability of a CP, but such movement may greatly degrade its C2 functions.

MAIN COMMAND POST SITE SELECTION

2-210. Units deploy their CPs on the battlefield according to METT-TC. Common methods include—

- CPs set up separately from unit locations.
- CPs set up with units.
- CPs using a combination of the above.

2-211. Setting up the CP independent of subordinate units separates the signatures associated with CP and helicopter operations. However, it makes face-to-face coordination more difficult unless adequate digital connectivity is available. Commanders decide which method to use during the IPB process.

2-212. The most important considerations for selecting any CP site are security and communications with higher, subordinate, and adjacent headquarters. Range of enemy artillery, accessibility to adequate entry and departure routes, cover, concealment, drainage, and space for dispersing are other considerations. An adequate LZ should be nearby. The S-3 selects the general location of the main CP, while the HHC commander and S-6 normally select the exact location. When selecting the general location of the CP, the S-3 chooses at least one alternate site in the event the primary site proves inadequate.

Offensive Operations

2-213. During offensive operations, the main CP should be well forward. In fast-moving operations, the main CP may have to operate on the move. Staff coordination and communications are usually degraded when CPs are moving; thus, CPs must train to operate while moving.

Defensive Operations

2-214. During defensive operations, the main CP normally locates in a secure location out of enemy direct and indirect fire capability to minimize its vulnerability. The exact location depends on the threat, terrain, road network, and ability to communicate.

Urban Operations

2-215. The main CP often sets up in built-up areas. Barns, garages, and warehouses minimize the need for detailed camouflage. Basements offer protection from enemy fires. Built-up areas also reduce IR and electromagnetic signatures.

Reverse Slopes

2-216. Reverse slopes cover and conceal CPs from direct observation and fires. Reverse slopes can degrade the enemy's ability to collect, monitor, and jam electronic transmissions. Electronic profiles run by the S-6 provide information for determining the ability to transmit and receive. Profile analysis by the S-2 provides information needed for determining the enemy's ability to degrade CP capabilities or intercept traffic.

Prominent Terrain Features

2-217. Units should avoid establishing CPs on prominent terrain or major road junctions. Such features are often enemy preplanned artillery and air targets.

MAIN COMMAND POST DISPLACEMENT

2-218. The main CP displaces in either a single or phased move. The method selected depends on METT-TC, the distance to be moved, and communications requirements. Movement degrades communication on all nets; however, the higher headquarters, brigade, and subordinate command nets must be maintained.

Maintaining contact with higher headquarters may require alternate communications means, such as aircraft or vehicle mounted systems. When operations are ongoing, moving the main CP is accomplished in a phased move in coordination with the displacement of the TAC CP. During displacement, critical aspects of C2 must be maintained. Displacements are planned ensuring the main CP is stationary during critical phases of planning and operations.

Displacement Steps

2-219. The S-3 establishes the general area for the new CP. The HHC commander, S-6, and a CBRN team conduct detailed reconnaissance. Following are steps for this displacement.

- The reconnaissance party identifies possible routes and sites. Locations must provide effective communications and accommodate all required vehicles and equipment. Several possible sites must be identified, reconnoitered, and planned to provide flexibility and alternate sites.
- The reconnaissance party makes route and site sketch maps showing the exact element locations within the new CP location.
- The S-3 or commander approves the primary and alternate sites.
- A movement order is published. An SOP practiced and drilled greatly reduces the effort required to produce the order.
- Security personnel and guides are dispatched. The security force ensures the area is clear of enemy and contamination, and the guides prevent wrong turns and assist elements in occupation. Signals are especially important for low visibility and night displacements.
- Reporting and coordinating functions are shifted as required. The shift may be within main CP echelons or to the TAC CP.
- CP and HHC elements prepare and execute movement per SOP. The main CP may displace in one echelon if the TAC CP can provide C2 for the interim. If the TAC CP cannot execute the required C2, the main CP displaces in two echelons. The first echelon displaces with enough assets and personnel to establish minimum C2. The second echelon remains in place and provides C2 until the first echelon assumes control, then displaces.

AUSTERITY

2-220. The main CP is a major source of electromagnetic and IR energy. Enemy forces may detect these emissions and fix the location of the main CP and place indirect fire, CAS, or electronic warfare (EW) strikes on it. The nature of the threat may require frequent CP movement. Frequent CP moves require the following considerations:

- The main CP should be as light as possible and drilled in rapid teardown, movement, and setup. The larger and more elaborate a CP, the less rapidly it can move.
- Frequent movement may hinder, C2 planning, degrade communications, and sacrifice time. Frequent moves may also increase the chances of enemy detection.

SECURITY AND DEFENSE

2-221. The HHC commander plans, organizes, and supervises the security and defense of the main CP. The plan establishes teams, squads, sections, and platoons. It also establishes a chain of command for perimeter defense and the quick reaction force (QRF). The brigade XO approves the plan. Refer to FM 3-04.126 and FM 1-113 for additional information on CP and AA procedures.

2-222. The threat is divided into three levels that provide a general description and categorization of threat activities, identify the defense requirements to counter them, and establish a common reference for planning guidelines. Table 2-3 lists these threat levels and their likely appropriate responses. Refer to FM 3-19.1 for additional information

Table 2-3. Threat levels

Threat Level	*Example*	*Response*
I	Agents, saboteurs, sympathizers, & terrorists.	Unit, base, & base-cluster self-defense measures.
II	Small tactical units, unconventional-warfare forces, guerillas, & bypassed enemy forces.	Self-defense measures & response forces w/supporting fires.
III	Large tactical-force operations (including airborne, heliborne, amphibious, infiltration, & bypassed enemy forces).	Timely commitment of a tactical combat force (TCF).

2-223. .Fighting positions are well prepared, mutually supporting, and known to all. Alarms are established and known to all; minimum alarms include ground, air, and CBRN attacks. Rehearsals are conducted and actions are greatly simplified if part of the SOP and drills are conducted often to ensure readiness.

2-224. The staff supports the HHC commander by providing personnel for defense and security. In an actual attack, the main CP continues C2 of the brigade unless the situation compels use of all personnel in the defense.

Reaction Forces

2-225. Reaction forces and attachments must be fully integrated into the overall plan. Each individual must have a clear and current SU of friendly and enemy forces in the AO. For example, a CP reaction force should know if military police are conducting mounted patrols near the CP. The overall reaction force plan must integrate those military police units or establish boundaries between the reaction force and the military police unit.

2-226. A clear chain of command and training supported by battle drills are essential for reaction force preparedness. They must assemble and be ready to fight in 10 minutes or less. Proper preparation includes the following:

- Alarms should be the same throughout the brigade, division, and corps. These alarms should be in the SOP.
- Reaction plans are rehearsed and executed on a routine basis. Prior to deployment and at in-country training centers, multiple integrated laser engagement system gear and live or blank ammunition supplemented by pyrotechnics should be used whenever possible to enhance training realism. The reaction to a night attack on the main CP must be second nature if the enemy force is to be repelled.

2-227. Each reaction force assembles based on an alarm or orders. Personnel move to a predetermined rally point, establish communications, and conduct operations as required to counter the threat.

Preparation for Security and Defense

2-228. Physical preparation for defense of the main CP site includes—

- Ensuring each Soldier is briefed, and has a copy and understands the ROE (for complicated ROE, it is often necessary to conduct situational training exercises to ensure understanding).
- Concealment including use of urban areas and camouflage.
- Cover for fighting positions and protective shelters.
- Vehicle revetments, transitory vehicle dismount points, and parking areas.
- Protective wire barriers.
- Prepared defensive positions.
- Prepared alternate and supplementary positions.

- Prepared routes for supply and evacuation.
- Minefields to cover avenues of approach, if approved for use. Adherence to correct procedures makes mine recovery less dangerous when it is time to displace. Minefields must be observed.
- Prepared sleep areas that are dug in or revetted to protect against enemy direct or indirect fires.
- Listening posts/observation posts (OPs) covering approaches to the main CP. These positions must be prepared to prevent visibility when approached from the front.
- Devices such as ground surveillance radar, personnel detection devices, and field expedients to enhance early warning of enemy approach or infiltration.
- Crew served weapons emplaced to cover suspected avenues of approach. Cleared fields of fire.
- Wire and directional antennas to prevent detection by enemy EW elements.
- Air and ground patrols to inhibit observation and attack of the main CP. Returning aircraft should be given patrol areas to survey before landing. Ground patrols should conduct reconnaissance as required to detect enemy observers or civilians who may be enemy informants.
- Daily stand-to to establish and maintain a combat-ready posture for combat operations on a recurring basis. Stand-to includes all steps and measures necessary to ensure maximum effectiveness of personnel, weapons, vehicles, aircraft, communications, and CBRN equipment. Units assume a posture during stand-to enabling them to commence combat operations immediately. Although stand-to is normally associated with begin morning nautical twilight, unit operations may dictate another time.

Standing Operating Procedures

2-229. CP organization, operations, and sustainment must be standardized in the SOP. Personnel associated with a CP must be completely knowledgeable of CP aspects. Training drills are essential for CP movement, setup, teardown, security, and operations. Drills to counter loss of critical personnel and equipment must be standardized and practiced day and night. Critical SOP items include—

- Personnel duties for each phase of CP operations and movement.
- Communications setup priorities—radio, wire, LAN, TI, and SATCOM.
- Critical friendly and enemy information reporting.
- Maintenance of maps and graphics.
- Maintenance of computers and automation equipment.
- Setup, teardown, and movement duties.
- Camouflage priorities.
- Light and noise discipline.
- Maintenance of generators.
- COMSEC changeover times.
- Maintenance of journals.
- CP security and admission procedures.

Chapter 3
Employment

This chapter addresses employment aspects for the aviation brigade. Factors common to all four types of operations—offensive, defensive, stability, and civil support—include planning considerations based on METT-TC and air-ground integration throughout the planning, preparation, and execution of a mission. Emerging battlefield elements affecting operations procedures include asymmetric threat, airspace deconfliction, CBRN weapons, and special environments.

SECTION I – GENERAL

Contents

3-1. Aviation brigades are tailored to execute operations supporting the ground force commander. The principal role of the brigade is to set conditions for the success of its units.

3-2. Modularity of forces means that aviation brigades may be tasked to support units from different divisions, such as teaming of a light aviation brigade with one or more HBCTs. Especially in cases such as this, early development of liaison and combined training is important.

TYPES OF OPERATIONS

3-3. Army forces conduct offensive, defensive, and stability operations as part of joint campaigns outside the U.S. Army forces conduct civil support operations inside the U.S. and its territories to support homeland security. Homeland security has two components–homeland defense and civil support. Homeland defense includes offense, defense, and stability missions to defend the homeland if threatened by hostile armed forces. Civil support missions include support to civil authorities and law enforcement; protection of military and civilian critical assets; and response and recovery. Types of Army operations are outlined in table 3-1 and tables 3-2 through 3-4 (page 3-2).

Table 3-1. Types of offensive operations

Offense	*Definition*
Movement to Contact	Employs movement to develop the situation, establish contact, or regain contact with the enemy.
	Search and Attack: Utilizes smaller, light maneuver units and attack reconnaissance or air assault forces in large area to destroy enemy forces or deny area to the enemy.
Attack	Destroys or defeats enemy forces through aerial firepower, mobility, and shock effect.
	CCA: Application of Army aviation into the close fight using integrated air-ground operations.
	Interdiction attack: Combines ground-based fires, attack aviation, unmanned systems, and joint assets to mass effects (beyond friendly forces in contact). Focused on key objectives, fleeting HVTs, and threats to friendly maneuver.

Table 3-1. Types of offensive operations

Offense	*Definition*
	Special Purpose: Special purpose attacks achieve objectives different from those of other attacks. Counterattacks are usually phases of a larger operation. Raids and ambushes are generally single-phased operations conducted by small units.
Exploitation	Follow-up of gains to take full advantage of success in battle.
Pursuit	An action against a retreating enemy force.

Table 3-2. Types of defensive operations

Defense	*Definition*
Mobile	Orients on defeat or destruction of the enemy force by allowing it to advance to a point where it is exposed to a decisive attack.
Area	Orients on denying the enemy designated terrain. Conducted to defend specified terrain, when the enemy enjoys a mobility advantage over the defending force, when well-defined avenues of approach exist, and the defending force has sufficient combat power to cover the likely enemy avenues of approach.
Retrograde (Delay)	Mission that trades space for time while retaining flexibility and freedom of action.
Retrograde (Withdrawal)	A planned, voluntary disengagement that anticipates enemy interference.
Retrograde (Retirement)	A force not in contact with the enemy moves away from the enemy.

Table 3-3. Types of stability operations

Stability Operations	*Definition*
Civil Security	Protecting the populace from external and internal threats.
Civil Control	Regulating the behavior and activity of individuals and groups to reduce the risk to individuals or groups to promote security.
Restore Essential Services	Providing the populace with essential services including life-saving medical care, the prevention of epidemic disease, provision of food and water, provision of emergency shelter from the elements, and the provision of basic sanitation.
Support to Governance	Providing societal control functions that include regulation of public activity, rule of law, taxation, maintenance of security, control and essential services, and normalizing means of succession of power.
Support to Economic and Infrastructure Development	Operations to provide direct and indirect military assistance to local, regional, and national economic and infrastructure development to provide an indigenous capacity for continued economic and infrastructure development.

Table 3-4. Types of civil support operations

Civil Support	*Definition*
Support Civil Law Enforcement	Providing support to local, state, and federal law enforcement officers. In extreme cases, when directed by the President, Regular ARFOR maintain law and order under martial law.
Provide Support in Response to Disaster	Providing C2, protection, and sustainment to government officials at all levels to support governance until these agencies are able to function without Army support.
Provide Other Support, as required	Conduct response to natural or man-made disaster in order to provide essential services to affected area. Essential services include rescue, emergency medical care, prevention of epidemic disease, provision of food and water, provision of emergency shelter from the elements, and the provision of basic sanitation.

PRINCIPLES

3-4. The aviation brigade's primary role is to set conditions for success by—

- Ensuring required C2 facilities are in place and operational.
- Ensuring SU (enemy and friendly).
- Ensuring necessary liaison to and from other organizations is in place.
- Coordinating brigade's movements and operations within the OE.
- Having necessary support and sustainment.

3-5. Aviation brigades have the organic capability of striking an enemy from multiple directions, either in support of BCTs or independently. Attack reconnaissance aircraft carry a combination of missiles, rockets, and guns to destroy HPTs, shield maneuver forces as they move out of contact, and enable shaping of the OE. In addition to traditional attack functions, the attack reconnaissance unit executes all functions of traditional air cavalry. As an armor killer, it is deadly against massed-moving targets, and is also effective against enemy FA, AD, communications, logistics units, and point targets (bunkers, caves, windows in buildings). The attack reconnaissance unit cannot occupy terrain; however, it can deny terrain for a limited period of time with direct and indirect fire.

3-6. Assault and GS aircraft provide organic capability for air assaults, aerial mine delivery, and sustainment operations in support of the aviation brigade or supported BCT. Assault and GS aircraft allow the commander to shape the battlefield utilizing air assaults to support seizure of key terrain or allow light forces to gain a maneuver advantage over enemy forces. They also provide the commander the ability to conduct BCOTM with aerial C2 support and A2C2S equipped platforms. In addition, the aviation brigade has the ability to provide air movement of sustainment assets, aeromedical evacuation, CASEVAC, and heavy lift support to the BCT.

COMMAND AND SUPPORT RELATIONSHIPS

3-7. Command and support relationships are fundamental to aviation operations. Table 3-5 depicts appropriate relationships and responsibilities.

Table 3-5. Command and support relationship to inherent responsibility

Inherent Responsibilities are—	*If Relationship is—*					
	Command				*Support*	
	Attached	*OPCON*	*Tactical control (TACON)*	*Assigned*	*DS*	*GS*
Has command relationship with—	Gaining unit	Gaining unit	Gaining unit	Parent unit	Parent unit	Parent unit
Task-organized by—	Gaining unit	Parent & gaining units: gaining unit may pass OPCON to lower HQ	Parent unit	Parent unit	Parent unit	Parent unit
Receives sustainment from—	Gaining unit	Parent unit	Parent unit	Parent unit	Parent unit	Parent unit
Assigned position or AO by—	Gaining unit	Gaining unit	Gaining unit	Gaining unit	Supported unit	Supported unit
Provides liaison to—	As required by gaining unit	As required by gaining unit	As required by gaining unit	As required by parent unit	Supported unit	As required by parent unit
Establishes/maintains communication with—	Unit to which attached	As required by gaining & parent units	As required by gaining & parent	As required by gaining units	Parent unit: supporte	As required by parent

Table 3-5. Command and support relationship to inherent responsibility

Inherent Responsibilities are—	*If Relationship is—*					
	Command				*Support*	
	Attached	*OPCON*	*Tactical control (TACON)*	*Assigned*	*DS*	*GS*
			units		d unit	unit
Priorities created by—	Gaining unit	Gaining unit	Gaining unit	Parent unit	Supporte d unit	Parent unit
Gaining unit can impose further command or support relationship of—	Attached; OPCON; TACON; GS; DS	OPCON; TACON; DS, GS	GS; DS	N/A	**	N/A
* In the North Atlantic Treaty Organization (NATO), the gaining unit may not task organize a multinational unit.						
** Commanders of units in DS may further assign support relationships between their subordinate units and elements of the supported unit after coordination with the supported commander.						

COMMAND RELATIONSHIPS

3-8. Command relationships are assigned, attached, OPCON, or TACON. A subordinate unit of the aviation brigade is attached only to a unit that can support its logistics needs. The aviation unit is placed OPCON or TACON when it is to be used for a specific mission, the effective time of the relationship is short, or the gaining unit is unable to provide logistics support. Normally, the parent headquarters retains control of the aviation unit.

3-9. Aviation units are traditionally OPCON or attached when operating outside the brigade. At the division level, units are placed OPCON to other units in support of ground operations. When operating as part of a TF or augmenting another unit for an extended period of time, the unit will be attached.

Assigned

3-10. Assigned is the placement of units or personnel in an organization where such placement is relatively permanent. The organization controls and administers the units or personnel for the primary function (or greater portion of the functions) of the unit or personnel.

Attached

3-11. Attached is the placement of units or personnel in an organization where such placement is relatively temporary. Subject to limitations imposed by the attachment order, the commander of the unit receiving the attachment provides sustainment support above its organic capability. Normally, the parent unit is responsible for transfers, promotion of personnel, nonjudicial punishment, courts-martial, and administrative actions.

Operational Control

3-12. OPCON is the authority to perform those functions of command over subordinate forces involving organizing and employing commands and forces, assigning tasks, designating objectives, and giving authoritative direction necessary to accomplish the mission. OPCON may be delegated and includes authoritative direction over all aspects of military operations and joint training necessary to accomplish missions assigned to the command. OPCON normally provides full authority to organize commands and forces and employ those forces as the commander considers necessary in accomplishing assigned missions. OPCON does not, in and of itself, include authoritative direction for logistics or matters of administration, discipline, internal organization, or unit training.

Tactical Control

3-13. TACON is the command authority limited to the detailed, and usually local, direction and control of movements or maneuvers necessary to accomplish missions or tasks assigned. TACON is inherent in OPCON. TACON may be delegated. TACON allows commanders to apply force and direct tactical use of logistics assets, but does not provide authority to change organizational structure or direct administrative and logistics support.

SUPPORT RELATIONSHIPS

3-14. The support relationships of utility and heavy helicopter assets are DS and GS. Specific definitions and missions follow.

Direct Support

3-15. DS refers to a mission requiring a force to support another specific force and authorizing it to answer directly to the supported force's request for assistance. A unit assigned a DS relationship retains its command relationship with its parent unit, but is positioned by and has priorities of support established by the supported unit. Assault and cargo helicopter units will often be placed in a DS role for air movement operations, particularly logistics movement. When operating in a DS role, these missions can be coordinated directly between the aviation unit and the supported unit.

General Support

3-16. GS is support given to the supported force as a whole and not to any particular subdivision thereof. For example, assault helicopter units assigned at theater and corps levels may be placed in GS to several units within the theater or corps. These units will receive missions from their parent headquarters based upon support priorities established by theater and corps commanders. When operating in a GS role, the supported unit must request aviation support from the appropriate headquarters (division Assistant Chief of Staff-Operations (G-3) for divisional aviation assets, corps G-3 for theater aviation assets).

SECTION II – ARMY AVIATION MISSIONS

3-17. Aviation forces normally operate as part of the combined arms team integrated from BCT level to theater level. Aviation is organized and equipped to support both joint and Army operations. An aviation TF supporting the BCT primarily conducts reconnaissance, security, CCA, air assault, air movement and aeromedical evacuation. When reinforced with ground combat units, the aviation brigade can be employed as a BCT.

RECONNAISSANCE

3-18. Attack reconnaissance aircraft are employed to support the commander's scheme of maneuver and significantly extend the OE of both the BCT and echelons above the BCT. Attack reconnaissance aircraft assist in locating the threat, building and sharing the COP, enhancing force protection, enabling freedom of movement, clearing the way for air assault missions, securing routes for aerial/ground resupply, and allowing the commander to focus combat power. Sensor video recording capability can provide the supported commander excellent reconnaissance and BDA information.

3-19. Attack reconnaissance assets can fight for information. They work through and counter enemy deception efforts, provide an expedient and reliable means of assessing terrain that the enemy is trying to configure to its advantage, can further develop the situation, and can effectively disseminate real-time information to commanders. The organic weapons systems of attack reconnaissance aircraft enhance the synergy achieved through employment of external fires and effects, giving commanders at all levels a robust counter reconnaissance capability.

Aerial Surveillance

3-20. Primarily a mission for UAS, the aviation brigade may conduct surveillance with manned and unmanned assets. Aerial surveillance is defined as systematic observation to obtain detailed information of a specific target or area. The focus of surveillance is generally a point target such as a house, car, section of road, or any other defined area with specific threat indicators to trigger priority intelligence requirements (PIRs). Brigade elements use the same fundamentals to conduct surveillance as with any reconnaissance mission. When performed by the ARB, surveillance is normally overt in nature with the purpose of deterring enemy movement or activity.

SECURITY

3-21. The aviation TF supporting the BCT can conduct security operations. Each can accomplish screen, guard, and cover operations with augmentation for the latter two operations. Security operations are particularly valuable during early entry operations when the COP is degraded and dynamics of the battlefield change fast. The combination of attack reconnaissance aircraft and UAS enable commanders at all levels to quickly move or deploy interactive and interpretive intelligence collectors over great distances to provide early warning, and gain and disseminate a timely picture of the battlefield. These aircraft quickly transition from a reconnaissance/counter reconnaissance or security mission to an economy of force or attack mission to provide reaction time, maneuver space, and protection for air-ground operations. The ACS, not normally a part of an aviation brigade, provides BCTs with added flexibility to conduct operations throughout their entire AO.

ATTACK

3-22. An attack is an offensive operation that destroys or defeats enemy forces, seizes and secures terrain, or both. Attack reconnaissance units of the aviation brigade conduct attacks in support of higher headquarters and supported BCT commanders. The ARB and ARS of the aviation brigade conduct both interdiction attacks and CCAs.

Interdiction Attack Operations

3-23. The interdiction attack capability of the aviation brigade, particularly when coupled with Army and joint fires and effects, extends the battle to the maximum range of organic and supporting sensors. The aviation brigade headquarters is most appropriate for the planning of interdiction attacks. During the conduct of interdiction attack operations, the aviation brigade normally requires OPCON or DS long range fires and supporting joint assets. Refer to FM 3-04.126 for additional information on interdiction attack operations.

Close Combat Attack

3-24. The ARB and ARS of the aviation brigade provide the commander with the capability to support ground forces that are engaged. CCA is carried out with direct fire weapons supported by indirect fire and CAS. The range between combatants may vary from several thousand meters to hand-to-hand combat. During CCA, attack reconnaissance aircraft engage targets near friendly forces, thereby requiring detailed integration of fire and maneuver of ground and aviation forces. To achieve desired effects and reduce risk of fratricide, air-ground integration must take place down to team levels. Refer to FM 3-04.126 for additional information on CCA operations.

MOVEMENT TO CONTACT

3-25. A movement to contact is conducted to gain initial contact or regain lost contact with enemy forces. The ARB, ARS, and ACS are ideally suited to conduct the movement to contact mission for the aviation brigade or supported BCT commander. Using joint, combined, and organic fires, the attack reconnaissance assets harass, impede, and destroy enemy elements to preclude their influence on the BCT main body.

Attack reconnaissance units support the ground forces with fires, maintain surveillance, and contain small forces until follow-on elements arrive to destroy the enemy.

SEARCH AND ATTACK

3-26. The search and attack mission utilizes smaller, light maneuver units and attack reconnaissance or air assault forces in large areas to destroy enemy forces or deny area to the enemy. Search and attack operations may be conducted against a dispersed enemy in close terrain unsuitable for ground maneuver, in sustainment areas against enemy SOFs or infiltrators, or as an area security mission to clear assigned zones. The search and attack technique is best used when the enemy is operating in small teams using hit-and-run tactics over a large area in a generally decentralized manner.

AIR ASSAULT

3-27. Aviation brigade assault and GS helicopter assets provide the maneuver commander the ability to conduct air assaults. Air assault operations for heavy BCTs are normally ground maneuver company sized air assaults. Air assault operations for forced entry operations and in support of infantry BCTs are ground maneuver battalion TF sized air assaults. Air assault operations extend the tactical reach of the maneuver commander, negate effects of terrain, seize key nodes, attain the advantage of surprise, and dislocate or isolate the enemy. The aviation brigade at division level has the organic capability to air assault the dismounted elements of a combined arms battalion and its required support equipment in a single lift and to provide air assault security.

AIR MOVEMENT

3-28. Air movement operations are conducted to reposition units, personnel, supplies, equipment, and other critical combat elements in support of current and/or future operations. The utility and cargo helicopters of the aviation brigade supplement ground transportation to help sustain continuous offensive and defensive operations. Air movement requires extensive precoordination with the supported force to ensure loads are properly rigged and placed in the PZ. FARPs emplaced by lift aircraft and ground assets enable aviation to support operations throughout the AO.

COMMAND AND CONTROL SUPPORT

3-29. The A2C2S, a UH-60-based package, represents a significant enhancement to the commander's ability to C2 forces. The A2C2S has the following five operational roles:

- BCOTM platform.
- Ground TAC CP.
- Jump TAC CP.
- Early entry CP.
- First responder during national disasters.

3-30. Onboard communications linkages allow the commander to be continuously in contact with committed forces, untethered to a ground based operations center, maintain SA, issue and receive FRAGOs with graphics, synchronize fires and maneuver, and extend his or her coverage throughout the entire OE. A2C2S systems are normally found in the GSAB of the aviation brigade. Refer to FM 1-113 for additional information.

AEROMEDICAL EVACUATION

3-31. Evacuation of casualties is the responsibility of the health support service (HSS) system. Air evacuation is the preferred method of evacuation of seriously wounded and ill Soldiers. The aviation brigade has an organic AAMC found in the GSAB. AAMC assets can collocate with HSS organizations, the aviation TF, the supported BCT, or higher to provide air ambulance support throughout the AO. Air ambulance aircraft are equipped with medical personnel and equipment enabling en route care of casualties.

CASUALTY EVACUATION

3-32. Assault and GS helicopter units may conduct CASEVAC operations when the number of medical aircraft is inadequate or not readily available. Refer to FM 1-113 for additional information.

PERSONNEL RECOVERY

3-33. All component commanders are responsible for establishing and coordinating recovery operations. For the BCT, the division and corps have additional communications linkages and detection capabilities, which may enable the rescue operation to be performed with greater safety and efficiency, within the constraints of METT-TC. The division and corps will then augment subordinate elements with the required assets in order to accomplish the PR mission. Corps PR is planned to support its own operations and to provide mutual PR support at both the intra- and inter-service levels as required. PR contingencies are incorporated into all mission plans; SPINS will be issued for each plan and brigades will be prepared to generate PR support requests.

3-34. Aviation brigade aircraft are not normally equipped, nor are personnel trained, to perform the CSAR mission. Appendix C provides additional information on PR operations.

AVIATION ENABLING MISSIONS

DOWNED AIRCRAFT RECOVERY

3-35. Downed aircraft recovery operations are coordinated at division and corps level by the ASB. The appropriate ABTF or GSAB of the aviation brigade will normally accomplish this mission. The goal is to recover aircraft with minimal risk to Soldiers involved in the operation.

AVIATION MAINTENANCE

3-36. The highly technical and complex aircraft of the aviation brigade require robust and redundant maintenance, repair, and ground support. Aviation maintenance must be fluid and able to adapt to the complexities and challenges of accelerated operating tempos (OPTEMPOs) and maximize the aviation brigade's contribution to the overall ground component commander's plan.

FORWARD ARMING AND REFUELING POINT OPERATIONS

3-37. Attack reconnaissance, utility, and cargo units all depend on the FARP to provide fuel and ammunition where and when they are needed. The FARP increases the time on station and extends the range of aircraft for by reducing the turnaround time associated with refueling and rearming. FARPs thereby give the commander more time to apply continuous pressure on the enemy. Refer to FM 3-04.104 for additional information.

AIR TRAFFIC SERVICES

3-38. ATS assets promote safe, flexible, and efficient use of airspace. They provide AC2 and ATS support enabling commanders to orchestrate air and ground maneuver, lethal and nonlethal fires, and AD to conduct decisive operation. ATS support is provided through automated airspace planning and en route services, terminal control tower, precision recovery, and airfield operations services throughout the BCT and division AO. These assets provide ATS and AC2 support through the tactical airspace integration system (TAIS) throughout the corps AO. TAIS is the AC2 node of the ABCS. AC2 cells organic to the battle staff at brigade and above assist in deconflicting, synchronizing and integrating all airspace requirements, including UAS. AC2 cells develop and maintain a real-time single integrated air picture through multi-path communications with all members of the air-ground team, allowing simultaneous access to airspace.

3-39. Army aviation can expect to operate from diverse locations often with aviation units from other services and nations. The desire of all aviation forces to operate from airfields or improved areas increases the probability of the brigade collocating with several different aviation units. This also increases the burden on ATS to deconflict operations, necessitating close coordination between commanders and the establishment of standard procedures. Refer to FM 3-04.120 for additional information on ATS operations.

Airfield Management

3-40. When more than one unit occupies an airfield, the appropriate joint forces commander will appoint an airfield commander. The airfield commander is responsible for the safe operation and accommodation of aircraft. This is accomplished through construction and maintenance of facilities and implementation of procedures and controls. Responsibility for daily operations can be delegated to an airfield manager.

3-41. Care must be taken at airfields to mitigate effects of environmental conditions on aviation operations. Severe weather and temperatures can cause catastrophic damage to equipment and loss of life, not to mention the corrosive effect of wind, rain, snow, and sand. Wartime OPTEMPO and consolidation of several units in one location can exacerbate wear and tear on airfield facilities and runways. Airfield managers must allocate engineer resources to reinforce, improve, and maintain high-traffic areas and those structures not originally designed to accommodate sustained operations.

SECTION III – AIR-GROUND INTEGRATION

3-42. Aviation and ground units require effective integration and synchronization to conduct operations successfully and minimize the potential for fratricide and civilian casualties. Integration starts at home station with the implementation of effective TACSOPs, habitual relationships, and training. It continues through planning, preparation, and execution of the operation.

3-43. To ensure effective integration, commanders and staffs must consider some fundamentals for air-ground integration. The following fundamentals provide the framework for enhancing the effectiveness of both air and ground maneuver assets:

- Understanding capabilities and limitations of each force.
- Use of SOPs.
- Habitual relationships.
- Regular training events.
- C2.
- Maximizing and concentrating effects of available assets.
- Employment methods.
- Synchronization.

3-44. Synchronization is the merging of the air and ground fights into one with the goal being proper application of aviation capabilities in accordance with the supported BCT commander's intent. Synchronization ideally begins early in the planning process with the involvement of the BAE. The BAE advises the BCT commander on aviation capabilities and the best way to utilize aviation to support mission objectives. Of equal importance is ensuring the LNO/BAE pass along task and purpose for aviation support and continually provide updates as needed. Simply stated, ensuring the aviation brigade and subordinate unit staffs fully understand the BCT scheme of maneuver and commander's intent is critical to successful air-ground integration.

3-45. Employment of attack reconnaissance aviation with ground maneuver forces requires coordinated force-oriented control measures and the CCA call for fire allowing aviation forces to support ground maneuver with direct fires while minimizing fratricide risks. LNOs should identify early in the planning process the minimum BCT graphics required for operations (boundaries, phase lines, attack by fire positions, objectives, etc.). BAE and LNO personnel should also ensure that supported units are familiar with CCA procedures and marking techniques discussed in FM 3-04.126.

COMMAND AND CONTROL

3-46. Aviation assets normally remain under aviation brigade, ABTF, or aviation battalion control. Subordinate battalion and company commanders operate on the command network but coordinate detailed actions on other nets or face-to-face. The commander ensures the focus of subordinate elements remains synchronized while executing various missions. He or she also clarifies coordination priorities and issues orders to each subordinate element, particularly on support issues, such as FARPs. This does not preclude direct coordination between ground and aviation elements.

AIR-GROUND CONTROL

3-47. Formation of air-ground TFs or teams may be used as an alternate C2 method. This relationship is used to deal with a specific situation on a temporary basis. OPCON is the normal command relationship. Specific employment guidelines must be established before operations. Air-ground teams are best used when decentralized company operations are required. Based on METT-TC, control may reside with either the ground or air commander. Rehearsals are essential with a focus on clearance of fires procedures.

SECTION IV – OPERATIONS IN ASYMMETRIC ENVIRONMENTS

3-48. Recent military conflicts have gravitated from conventional engagements executed on a contiguous battlefield to more undefined operations against an asymmetrical threat on a noncontiguous battlefield. Although a conventional, contiguous environment is still relevant, commanders must be familiar with the emerging unconventional battlefield environment.

ASYMMETRY

3-49. Asymmetry is defined as the dissimilarities in organization, equipment, doctrine, capabilities, and values between other armed forces (formally organized or not) and U.S. forces. Asymmetric operations exist when forces, technologies, and weapons are significantly different, or a resort to terrorism and rejection of more conventional ROE are normal. Asymmetric engagements can be extremely lethal, especially if the target is not ready to defend itself against the asymmetric threat. Asymmetry tends to decay over time as adversaries adapt to dissimilarities exposed in action. The likelihood of asymmetric attack increases with the continued conventional dominance of U.S. forces and the growing threat of weapons of mass destruction (WMD).

3-50. Countering asymmetric attacks may require altering ROE, organization, doctrine, training, or equipment. To reduce the vulnerability of asymmetric attacks and minimize their effects, Army organizations, training, and equipment emphasize flexible employment in diverse situations. Protective measures, such as physical security and operations security (OPSEC), lessen the effects of asymmetry. A credible CBRN defense capability at the tactical level deters the use of WMD. The threat of asymmetric action requires emphasis on security, even in low-threat environments.

THREAT

3-51. Often unable to challenge the Army in conventional combat, adversaries seek to frustrate Army operations by resorting to asymmetric means, weapons, or tactics. Attacks pose threats from a variety of directions with a broad range of weapons systems designed to stress the enemy's defenses. For example, luring attack helicopters into an AD artillery ambush by displaying a prominent target (tank) is a common asymmetric operation. The enemy can also be expected to take refuge in any available restrictive or urban terrain to conduct operations.

3-52. Potential threats vary from heavy conventional units to adaptive, asymmetric forces structured for local and regional use. Enemy forces may be widely dispersed and numerically superior. In nontraditional environments, the enemy can be expected to take advantage of restrictive and urban terrain. Adversaries will also seek and obtain technologies challenging U.S. strengths in information technology, navigation, night vision systems, and precision targeting and strike capabilities. The proliferation of WMD and long-range delivery systems will enable adversaries to threaten U.S. forces at greater ranges with increased lethality and precision.

3-53. Because of the difficulty in predicting asymmetric threats, IPB is essential. Accurate intelligence decreases the uncertainty critical to enemy success. Operational success requires identifying enemy capabilities (strengths and vulnerabilities), intentions, and COAs. Identifying and disseminating intelligence gaps to operational units prevents a false sense of security.

3-54. Army aviation primarily utilizes reconnaissance, search and attack, and CCA closely integrated with ground maneuver elements. Without a massed threat, friendly units are organized into small, decentralized, combined arms teams.

3-55. Aviation can expect to conduct 24-hour operations supporting reconnaissance, CCA, QRF, and resupply requests. The brigade will depend on each BCT BAE and aviation battalion liaison teams working with supported units to coordinate aviation support requests.

3-56. Small arms, rocket-propelled grenades, and shoulder-fired surface-to-air missiles creates a dangerous environment for rotary-wing aircraft. By operating in small teams, engaged aircraft can focus on survivability while directing lethal fires from CAS or attack aircraft on enemy positions.

3-57. Because of the constantly adapting enemy, aviation brigade elements should minimize steady state trends. Alternating routes, take off times, and TTPs may reduce the threat generated by an asymmetric enemy. Running fire, suppressive fires, and maneuvering flight are most effective during CCA or unexpected en route engagements.

TRAINING

3-58. Training and preparation are critical to countering the asymmetric threat and minimizing the inherent advantages. Collective training in the AO, or in similar conditions prior to deployment, will promote familiarity with the terrain and enemy tactics. Aviation units use aviation combined arms tactical trainers (AVCATT) and aviation training exercises (ATXs) with asymmetric terrain databases to develop and validate TTP for use in projected environments. By minimizing the element of surprise and maintaining heightened security, enemy operations are more effectively countered.

3-59. The present conventional dominance of the U.S. military and recent conflicts imply future operations are more likely to be asymmetric. Adhering to the "train as you fight" philosophy, training must involve more asymmetric scenarios and continue once deployed in that environment. Since each situation is unique, units must experiment and adapt to their specific environment disseminating and training effective TTP.

FRATRICIDE

3-60. The potential for fratricide increases due to the fluid nature of the noncontiguous battlefield and the changing disposition of attacking and defending forces. The presence of noncombatants in the AO further complicates operations. In this setting, commanders exercise judgment in clearing fires, both direct and indirect.

SUSTAINMENT

3-61. The dispersed nature of noncontiguous AOs often separate flight, maintenance, and refuel operations, requiring extended LOCs and innovative means to conduct sustaining operations. Demand for helicopter security of convoy and air movement operations can be expected to increase as sustainment distances and asymmetric threats increase.

COMMAND AND CONTROL

3-62. Noncontiguous and asymmetric operations frequently involve a larger AO and increased communication requirements. Reliable C2 architecture is critical to aviation's responsiveness.

SECTION V – CHEMICAL, BIOLOGICAL, RADIOLOGICAL, AND NUCLEAR

3-63. CBRN functions are found at division level and below; chemical, biological, radiological, nuclear, and high yield explosives (CBRNE) functions are found at corps level and higher. U.S. forces are most likely to encounter a CBRN environment, especially when facing a militarily less-capable threat resorting to asymmetric responses. The aviation brigade must avoid the effects of CBRN weapons, take protective measures, respond to their use, and continue the mission. SOPs and training are the best preparation for operations in a CBRN environment. Refer to FM 3-11 and FM 3-11.4 for additional information.

3-64. The commander must consider exposure guidance from higher headquarters, enemy capability, unit mission, and condition of the unit when establishing the unit's mission-oriented protective posture (MOPP). Because of the degradation of aircrew effectiveness in MOPP equipment, intensive fighter management is required. To reduce risk in a CBRN environment, units must—

- Avoid detection.
- Retain mobility.
- Seek terrain shielding by carefully selecting AAs and preparing shelters and fighting positions.
- Instill discipline and physical conditioning to prepare Soldiers for the confusion and physical demands of a CBRN environment.
- Plan for continued operations if attacked.

CONTAMINATION AVOIDANCE

3-65. The term avoidance does not necessarily mean aborting a mission or suspending operations. Soldiers enter contaminated areas only when necessary; however, it is preferable to avoid these areas by bypassing when possible. Units use the CBRN warning and reporting system, and reconnaissance, monitoring, and surveys to assist in identifying contaminated areas.

PROTECTIVE MEASURES

3-66. When elements cannot avoid contamination, or are under direct attack, Soldiers must take appropriate actions to survive. Specific actions are taken before, during, and after attack. To sustain operations in CBRN environments, personnel must understand and practice individual and collective protection. Individual protection involves those measures each Soldier must take to survive and continue the mission. These measures include immediately donning MOPP gear, seeking cover, and using other protective equipment and devices. Collective protection provides a contamination-free environment for selected personnel and precludes continuous wear of MOPP gear. Considerations for CBRN protection include—

- Positioning CBRN reconnaissance assets at likely locations for enemy employment.
- Combining reachback intelligence with battlefield sources to anticipate enemy use of WMD.
- Using smoke to support disengagement.

SECTION VI – SPECIAL ENVIRONMENTS

3-67. The brigade will be called upon to execute its mission in a variety of environments. It is imperative that commanders understand the impact of these environments on their Soldiers and equipment. Commanders need to think through the impact of environmental conditions and provide necessary training. The Army's concept of "just in time" training, supported by the use of distance learning products, provides opportunities for commanders to meet some of the unique training challenges special environments demand.

URBAN ENVIRONMENT

3-68. In urban areas, fields of fire are restricted, landing areas are limited, and buildings provide cover for enemy forces to engage helicopters with near impunity. The presence of noncombatants, protected structures, and important resources and facilities normally demands careful weapons and munitions selection to avoid collateral damage. The proximity of enemy and friendly ground forces increases the risk of fratricide. Communications may be degraded by many structures. Thermal effects from paved surfaces and the channeling effects of buildings can cause wind conditions to vary significantly from point to point. Special, restrictive ROE should be expected. Standoff is key to aviation survival.

3-69. Manmade structures and the density of noncombatants in urbanized terrain affect the tactical options available to commanders and aircrews. Whether engaged in major theater war or stability operations, the aviation brigade will conduct operations in urbanized terrain. This is partly due to growing populations, but also results from a potential adversary's tendency to create a noncontiguous battlefield rather than attempt to face U.S. forces directly. Potential adversaries can be expected to use urbanized terrain for cover and concealment, and reduce U.S. combat superiority by taking advantage of weapons restrictions and reduced options available to commanders under ROEs, rules of interaction (ROIs), and Laws of War. ROE and ROI must be rehearsed, practiced, and reinforced continually throughout the operation. FM 3-06.1, FM 1-113, and FM 3-04.126 provide additional information on operations in urban terrain.

CONDUCTING OPERATIONS

3-70. U.S. forces may conduct operations in urbanized terrain for the following reasons:

- The unit is force-oriented and the enemy occupies a built-up area.
- The political importance of the urban area justifies using time and resources to liberate it.
- The area controls key routes of commerce and provides a tactical advantage to the commander controlling it.
- The enemy in the urban area, if bypassed, might be able to interdict LOCs.
- Critical facilities within the urban areas must be retained or protected.

3-71. U.S. forces may avoid operations in urbanized terrain for the following reasons:

- The enemy, if bypassed, presents no substantial threat to friendly operations.
- The commander does not have sufficient forces to seize and clear the area.
- The urban area is declared an open city, making an attack illegal under the Law of War.

PLANNING AND EXECUTION

3-72. Operations in urban terrain generally follow the same planning and execution concepts as in other terrain; however, special planning and consideration of characteristics unique to urban terrain are required. Refer to FM 3-06.1 and FM 3-06.11 for additional information on urban operations. Aircraft must standoff to engage targets in urban areas. Overflight and engagement of targets within urban areas may require night operations and special preparation due to possible enemy direct fire at close range. Hovering in urban areas exposes aircraft to small arms fires and should only be attempted if essential to the mission and adequate overwatch fires are available. Wire, tower, and antenna hazards are especially prevalent and must be considered in the IPB. Other examples include—

- Demographics of the local population.
- Subterranean, ground level, and above-ground terrain analysis.
- Civilian maps and diagrams.
- Airfields, helipads, and rooftops used as LZs.
- Structures and areas protected by the Law of War or restricted by ROE.
- Supplementary electronic and visual signals to differentiate friend from foe.
- Weapons selected to produce the desired affect while minimizing collateral damage and maximizing standoff.

3-73. Helicopters can emplace forces on rooftops, in parks, stadiums, parking areas, and other similar areas. The presence of wires, poles, antennas, debris, and other obstacles may limit some landing areas. Attack reconnaissance aircraft cover landings and minimize exposure by engaging targets using running fire and diving fire. Helicopters must minimize ground time and hovering to avoid sniper, grenade, and rocket-propelled grenade engagement when inserting or overwatching forces.

3-74. Because of the dynamics of urban growth, current maps and photographs are essential for accurate planning. In the absence of these materials, detailed reconnaissance is required to minimize risk.

CIVIL CONCERNS

3-75. Operations in urbanized terrain have a significant impact on noncombatants. Special considerations are required. Units should maintain liaison with local police, ATS, and civil and military authorities.

Care of Civilians

3-76. Civilians may be removed from the area or protected in their homes. In some cases, the aviation brigade may be required to arrange for supply, transportation, medical care, and other support for civilians.

Security

3-77. The threat of espionage, sabotage, and terrorism must be carefully considered and guarded against during all phases of aviation operations.

Civilian Interference with Military Operations

3-78. The aviation brigade must ensure civilians do not interfere with the execution of military operations. The aviation brigade relies on military police, Staff Judge Advocate representatives, and human intelligence teams to liaison with local law enforcement officials. They control displaced civilian flow while they help identify and interrogate any suspicious displaced persons moving through the AO.

MOUNTAINS AND HIGH ALTITUDES

3-79. Mountainous environments—particularly the severe and rapidly changing weather—affect aircraft performance capabilities, accelerate crew fatigue, and influence basic flight techniques. Limited visibility operations in the mountains are extremely hazardous and require extensive aircrew training. Common problems associated with mountain operations are more complex at night, even when using night vision devices (NVDs).

3-80. While high altitude limits load-carrying capabilities, compartmentalized mountain terrain enhances rapid movement to the flanks and rear of an isolated enemy force. Enemy mechanized forces are slowed and canalized as they move up steep grades and down narrow valleys or are restricted to roads and trails. Mountains provide excellent terrain masking and allow easy avoidance of radar and visual acquisition; however, high ridges also provide effective FPs for AD guns and hand-held missiles. Mountain flying techniques are critical to taking advantage of this terrain.

HIGH-ALTITUDE TRAINING SITE

3-81. The High-Altitude Army Aviation Training Site, located in Eagle, Colorado, provides excellent high altitude and power management training for rotary-wing aviators. If possible, all pilots-in-command should attend this course before deploying. The course is valuable for operating at high gross weights or high altitude. Course length is 1 week and an exportable training package is available.

SNOW, ICE, AND EXTREME COLD WEATHER

3-82. Operations in snow, ice, and extreme cold weather pose operational and maintenance challenges. Unpredictable weather conditions complicate the planning process and commonly cause a deliberate reduction in OPTEMPO.

3-83. Ice can prevent proper weapons and missile function. Uncovered aircraft require frequent checks and services to prevent icing. Aircraft that become ice-covered may take hours to de-ice. Aircraft skis may be required.

3-84. The depth and consistency of snow in a landing area is an important consideration. Blowing snow can create whiteout conditions, especially during takeoff, landing, or hovering. The AMC should increase the spacing between aircraft in snow-covered landing areas, and preparation teams should pack the snow prior to arrival. Low and slow-flying aircraft may produce large snow clouds the enemy can easily detect. They may also blow snow off trees leaving a trail visible to enemy aircrews or UAS.

3-85. Frozen bodies of water make excellent LZs as they are level and have minimal loose snow due to wind scouring. Suitability of the landing area depends on aircraft weight and ice thickness.

3-86. Navigation using terrain following and maps is degraded over snowfall and frozen waterways. Navigational aids (NAVAIDs) and global positioning systems (GPSs) are essential in this environment.

3-87. Units not normally operating in these conditions should request SOPs and guidance from those units experienced in these conditions. Measures to combat lower temperatures and snow will constrain the OPTEMPO.

JUNGLES

3-88. Dense jungles and wooded areas degrade fields of fire and target identification, and can negate advantages afforded by superior acquisition systems. Humid, tropical air decreases the effectiveness of optics. It also decreases payload capacity. While a tropical jungle can be one of the harshest terrains available for aviation operations, mobility advantages offered by aviation over ground forces are exponentially increased.

3-89. Downed aircraft without a smoke signature can be difficult to locate. Aviation life support system radios, GPSs, and survival gear are especially critical as are effective flight following use of GPS coordinates and preplanned posted routes. SOPs must address aircrew recovery.

DESERTS

3-90. The brigade can effectively operate in the desert, but open desert terrain increases the unit's vulnerability to enemy long-range observation and acquisition. Leaders should take advantage of periods of limited visibility, or consider a wider dispersion of aircraft.

3-91. The weather in desert regions can be extremely unpredictable. Sandstorms, accompanied by constantly fluctuating wind speeds, may reduce visibility from more than 50 kilometers to zero in less than 5 minutes. Pilots must be carefully briefed on prevailing weather conditions before takeoff. Warning of any expected variations in conditions must be transmitted immediately to all airborne aircraft.

3-92. Desert surface composition affects the choice of LZs, maintenance sites, FARPs, and operating bases. Hard-packed sand provides the best conditions; however, prolonged use will produce finer sand particles resulting in degraded ground and air operations. Leaders must seek airfields and hardstand surfaces when possible; if unavailable, sealant, oil, diesel fuel, or water may be applied after a thorough environmental assessment to limit dust clouds.

3-93. Heat limits the load bearing capability of aircraft. Placing FARPs closer to objective areas can mitigate the effects of reduced payload capabilities. Aircrews can employ running landings to carry a greater payload. Because many deserts have extremes in temperature, missions are best conducted at night when temperatures are cooler.

3-94. Flight below 50 feet above ground level in a desert environment can be a difficult transition for brigade aircrews. Many aircrews use the IR searchlight to improve terrain definition while using night vision imaging systems. Units must balance the risk of sophisticated enemies detecting such searchlights against that of radar AD engagement at higher altitudes, or accidental terrain contact. Regular training with NVDs can reduce reliance on the IR searchlight and its accompanying risks.

3-95. Aircraft flying low and slow during takeoff, hover, and landing produce large dust clouds the enemy can easily detect. Dust clouds produce brownout conditions that obscure pilot vision during the day and under NVDs. These activities are extremely damaging to turbine engines, rotor blades, and nearby ground equipment, reducing their operational lifespan. Aircrews must minimize hovering, expedite takeoffs and landings, or fly instrument meteorological condition (IMC) if brownout occurs. Units must train in a desert combat environment to be comfortable operating a blacked-out aircraft in brownout conditions.

SHIPBOARD AND OVERWATER OPERATIONS

3-96. Shipboard operations provide many options to joint force and component commanders. Army helicopter operational capabilities are greatly expanded when ships are available for operations near large bodies of water and islands. Shipboard operations require special training prior to helicopters landing on or operating from ships.

3-97. Overwater operations may be necessary to defeat enemy waterborne operations or move from one location to another. As in desert environments, openness increases the unit's vulnerability to enemy long-range observation and acquisition. The lack of NAVAIDs and prominent terrain features makes navigation extremely difficult without GPS, Doppler, or some other form of navigation assistance. Overwater operations require special equipment and training (water wings, rafts, and helicopter emergency egress devices). Units normally not operating in these conditions should request SOPs and guidance from those units with experience in these conditions. Refer to AR 95-1and JP 3-04.1, for additional information.

SMOKE AND OBSCURANTS

3-98. Smoke and obscurants are integral parts of most potential adversaries' doctrine, tactics, equipment, and training. Enemy forces will use smoke to increase effectiveness and reduce vulnerability. Specifically, the enemy can use smoke to—

- Deny information.
- Mask use of chemical weapons.
- Disrupt movement, operations, and C2.
- Restrict NOE and contour flight.
- Reduce effectiveness of sensors, range finders, target designators, and visual observation.

OBSCURANT EMPLOYMENT

3-99. Through the use of smoke, the brigade can—

- Suppress visually sighted enemy AD systems and small arms.
- Sector portions of EAs, isolating part of the enemy force.
- Obscure LZ or PZ operations from enemy view.
- Screen displacement of attack reconnaissance aircraft while they move or break contact.

3-100. For deliberate operations, battalions can employ multi-spectral smoke-generating equipment. Helicopters can employ white phosphorus rockets on enemy positions to obscure vision if—

- Adequate numbers of rockets are available.
- Weather conditions are favorable.
- The mission is coordinated in advance with friendly forces in the immediate area.

3-101. The downside of friendly or enemy use of obscurants is a degraded performance of sensors and a potential negative effect on use of semi-active laser (SAL) Hellfire (radar frequency Hellfire is unaffected by smoke). During air-ground integration planning, both air and ground units must plan schemes of maneuver and support by fire positions that consider the effect smoke may have in obscuring friendly observation and designation.

SECTION VII – UNMANNED AIRCRAFT SYSTEMS OPERATIONS

3-102. UAS linked to brigade assets enhance operations. Maximum use of UAS and joint UAS assets can greatly reduce requirements on the commander's internal resources for security. UAS units can perform all basic observation tasks, thus freeing helicopters for higher priority actions. UAS integration can reduce flying-hour requirements and support fighter management. While TTP governing UAS operations are emerging, every opportunity to use UAS should be exploited. Refer to FMI 3-04.155 and FM 3-04.15 for additional information on UAS employment.

3-103. Communications and coordination with UAS controllers are essential for integrating UAS. If a UAS unit conducts screen of an area, accepts handover from or handover to an attack reconnaissance unit, necessary C2 must be planned in great detail to ensure proper coverage of the security area. The C2 of UAS and manned aircraft is further complicated if the unit controlling the UAS is at another location or the higher headquarters location.

3-104. Combined UAS and attack reconnaissance operations are an excellent force multiplier. SOPs, battle drills, rehearsals, and training exercises contribute to the success of combined UAS and manned aircraft system operations.

RECONNAISSANCE, SURVEILLANCE, AND TARGET ACQUISITION OPERATIONS

3-105. UAS capabilities make them ideal to support brigade reconnaissance and security missions. Locating enemy systems is a critical mission for UAS. UAS can cue brigade forces during screen, guard, and cover missions. Likewise, during economy of force missions, UAS can alert dispersed brigade forces to mass effects on a particular enemy force. The fielding/retro-fitting of armed UAS is a near future combat multiplier for the aviation brigade.

CONCEPTS OF UNMANNED AIRCRAFT SYSTEM AND AVIATION BRIGADE COOPERATIVE EMPLOYMENT

3-106. Three employment options for Army aviation and UAS assets are discussed below. These employment methods refer to tactical level UAS (Hunter, Shadow, and Warrior) that operate above the coordinating altitude. Small UAS such as Raven operate at the same altitudes as manned aircraft making airspace coordination and deconfliction problematic. See Appendix B of FMI 3-04.155 for more information on small UAS and manned aircraft deconfliction.

UNMANNED AIRCRAFT SYSTEM TO AVIATION UNIT HANDOVER

3-107. The staff section controlling UAS acquires the enemy force and maintains observation. UAS capabilities enable commanders to view the OE in near real-time. HPTs are handed over to the brigade for continued observation or destruction (figure 3-1, page 3-18).

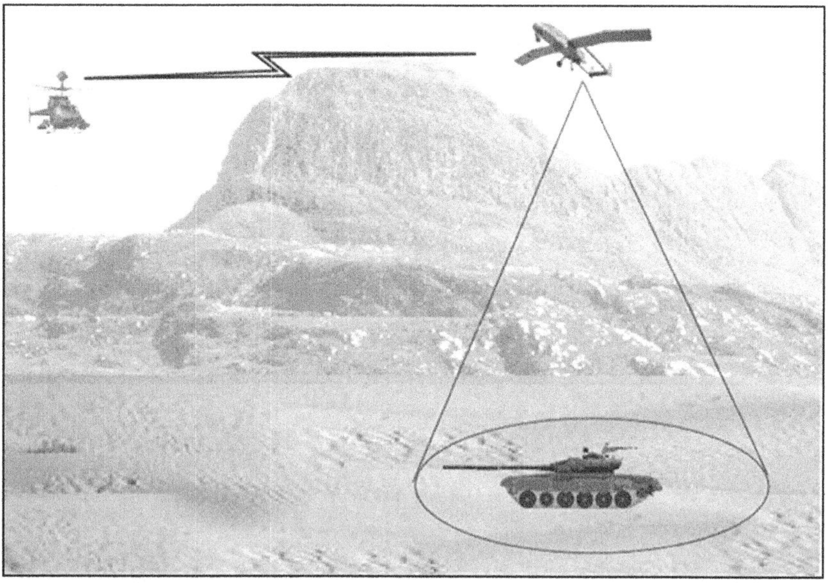

Figure 3-1. Unmanned aircraft system to aviation unit handover

3-108. This includes UAS equipped with laser designators teaming with attack reconnaissance helicopters to attack point targets with Hellfire missiles. This option allows for accurate and responsive fire while maintaining maximum standoff for manned aviation (figure 3-2, page 3-19).

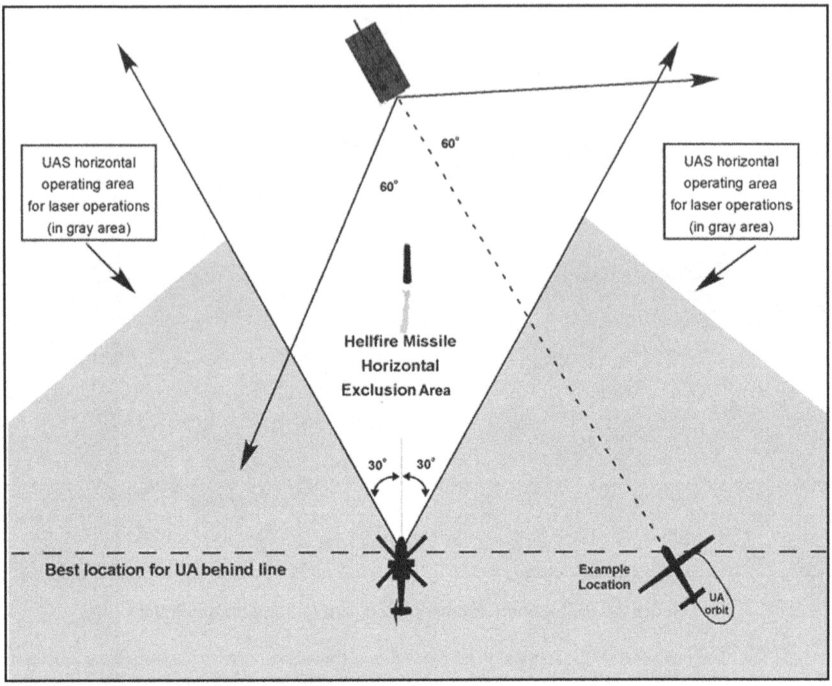

Figure 3-2. Unmanned aircraft system support to Hellfire (horizontal)

AVIATION UNIT TO UNMANNED AIRCRAFT SYSTEM HANDOVER

3-109. The brigade acquires the enemy force and maintains observation. HPTs are handed over to UAS for continued observation and engagement by FA or CAS. The brigade then conducts a bypass of enemy forces, and continues the reconnaissance effort or moves back to an FAA or the AA (figure 3-3, page 3-20).

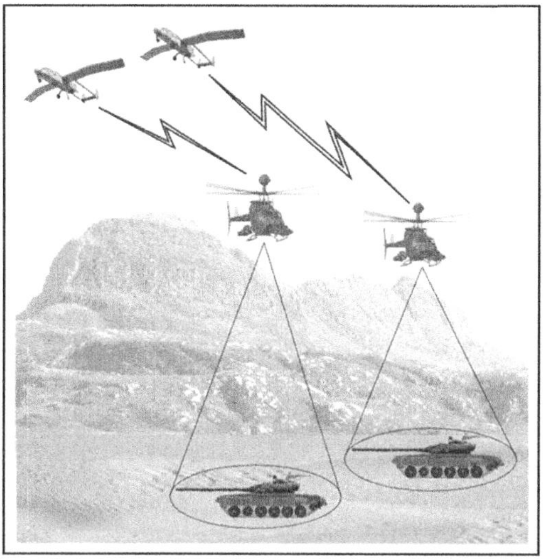

Figure 3-3. Aviation unit to unmanned aircraft system handover

Aviation Unit and Unmanned Aircraft System Area of Operations

3-110. METT-TC is the basis for aviation unit and UAS AO assignments. If the situation dictates, they can switch AO. This option maximizes the capabilities of both systems; however, it requires the most coordination. This option allows the brigade to extend its AO and concentrate manned elements on the most critical AO (figure 3-4).

Figure 3-4. Aviation unit and unmanned aircraft system area of operations

Chapter 4
Aviation Battalion Task Force Operations

The concepts expressed in this chapter are focused on formation and employment of ABTFs. These concepts and considerations also apply to the aviation brigade and should be reviewed and considered when the brigade receives additional aviation units.

SECTION I – GENERAL

Contents

4-1. Aviation brigades by design are intented to operate as TFs and create ABTFs based on METT-TC.

4-2. In order to provide appropriate forces for a particular mission or to cover large AOs, the aviation brigade can form ABTFs. The CAB divides its organic attack reconnaissance, utility, cargo, sustainment, and C2 assets to meet the requirments of the mission. Additional augmentation may be required in some cases. While the GSAB is capable of serving as an ABTF headquarters, it normally provides C2, cargo helicopter, aeromedical evacuation, and ATS assets to support brigade and ABTF operations.

4-3. The ABTF deploys with all personnel and equipment required to accomplish its mission. Competing requirements often challenge TF needs for limited equipment and personnel. TF shortfalls not met within the parent unit may have to be filled from other units. Additionally, internal and external ABTF operating procedures have to be tailored for combined or multinational operations and specific theater requirements.

FIXED BASE OPERATIONS

4-4. Deploying aviation units initially seek operating locations that best replicate their home station facilities and capabilities. The goal of this initial staging base is establishing and maintaining a secure area for aviation operations. If available, aviation units occupy areas in and around an airfield or improved surface to facilitate sustainment operations in preparation for combat. Desired facilities include an operational tower, NAVAIDs, hanger facilities, helicopter parking areas, and barracks. As the unit moves forward, this base may be used to conduct phases and depot-level maintenance.

SPLIT-BASED OPERATIONS

4-5. The aviation brigade can conduct split-based operations, defined as the division of logistics, staff, management, and command functions over two or more AOs.

4-6. By task organizing, the aviation brigade can realistically produce three ABTFs. If three ABTFs are deployed apart from the brigade, the aviation brigade is no longer capable of executing its core competencies.

OPERATIONAL OVERVIEW

4-7. The ABTF is assigned, attached, or OPCON to the supported unit. It may be deployed without a senior aviation headquarters or as part of a larger aviation force. The ABTF or its higher aviation headquarters normally operates within the AO of a supported BCT, division, corps, or joint TF. Elements of the ABTF may operate in multiple locations within that AO.

COMMAND RELATIONSHIPS

IN-THEATER

4-8. All aspects of planned command relationships must be considered to ensure necessary C2, liaison, and support personnel and equipment are provided to the ABTF organization. The following are examples of ABTF command relationship options. This list is not all inclusive; some of the relationships reflect emerging doctrine. The ABTF can be in a command relationship that is—

- Directly subordinate to the joint task force (JTF) headquarters (figure 4-1).
- Directly subordinate to the Army Component Command, s Component Command, or Naval Component Command.
- Further assigned, attached, or placed OPCON or TACON to any of the components above.
- Subordinate to a corps, division, or BCT.
- Subordinate to an aviation brigade.

In small scale contingencies, the ABTF may be assigned to a JTF. The JTF assigns the ABTF to a carrier strike group for the first phase of the operation. The carrier battle group places the ABTF TACON to the Marine air ground TF. Once the Marines achieve the objective, they are relieved by an Army division TF that will also become the land component commander. The ABTF is relieved of assignment to the carrier strike group and assigned to a BCT or the division's aviation brigade, if present.
In addition to normal planning associated with operations, the ABTF develops its OPLANs at its own headquarters as an integral part of its higher headquarters staff, or both.

Figure 4-1. Example of command relationship scenario

SECTION II – ORGANIZATION AND MISSION

4-9. The ABTF performs maneuver, support, and sustainment missions. It also possesses the unique ability of a single battalion-size organization to perform reconnaissance, security, attack, air assault, air movement, and C2 operations.

4-10. The following principles pertain to ABTF formulation:

- Aviation planning principles remain consistent across aircraft types and missions.
- Each organization combines its various TTP at TF level to provide a TF battalion-sized unit supplying aviation support across the full spectrum of operations.
- Once formed and trained, the ABTF possesses rapid deployment capability and conducts Army or joint operations in any OE.
- An ABTF may incorporate active Army and RC units.
- Proper ABTF organization, planning, and training are essential to accomplish assigned missions.
- Aviation maintenance and logistics support requires special emphasis.

ORGANIZATION

4-11. The ABTF is not a standing unit with a published TOE. An ABTF's organizational structure is tailored to meet mission requirements. There are numerous possibilities when organizing an ABTF. METT-TCs are driving factors for ABTF organization. Figure 4-2 provides a sample ABTF (heavy) organization.

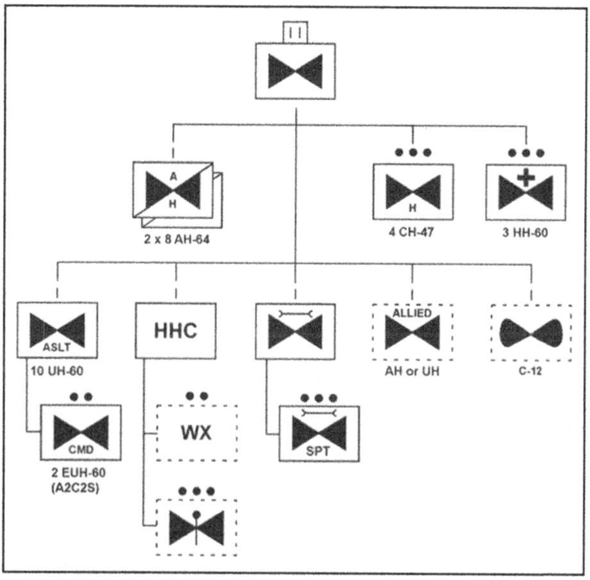

Figure 4-2. Sample aviation battalion task force (heavy) organization

4-12. The ABTF may receive additional assault and cargo helicopter forces from its divisional aviation brigade GSAB, theater aviation brigade, or uncommitted divisions to augment its air assault capability. Likewise, it may receive additional attack reconnaissance assets if employment is expected to be more reconnaissance, security, or attack centric. Such augmentation could be in the form of full companies, platoons, or sections, or it could be as small as additional crews with or without additional aircraft.

MISSION

4-13. An ABTF is capable of conducting all aviation missions. Specific ABTF tasks and roles include—

- Conduct screen operations to maintain surveillance and provide early warning of contact with enemy forces.
- Conduct guard and covering force operations as part of a larger force.
- Employ attack reconnaissance helicopter elements as part of a combined arms force of the respective higher headquarters.
- Conduct or support raids.
- Increase the tempo of friendly operations.
- Conduct operations to destroy enemy formations, communications, and logistics assets.

- Provide aerial escort and suppressive fires in support of air assault, convoy security and other TACOPS.
- Provide mobile firepower to exploit the effects of artillery and other indirect fire.
- Conduct limited J-SEAD operations.
- Coordinate and adjust indirect fires.
- Conduct JAAT operations with CAS and FA forces.
- Conduct company-sized air assaults (if seats are removed from UH-60s).
- Conduct fast-rope insertion/extraction system (FRIES)/special patrol infiltration/exfiltration system (SPIES) operations.
- Conduct air movement of supplies using external or internal loads.
- Conduct aeromedical evacuation, CASEVAC and personnel replacement operations.
- Provide refuel capability using Fat Hawk (external fuel tank refueling operations).
- Conduct aerial mine delivery operations (Volcano).
- Conduct C2 operations.
- Conduct DART operations.
- Participate in PR operations.
- Conduct operations in multiple locations simultaneously.
- Conduct day and night aviation operations during visual and marginal weather conditions.
- Conduct limited operations during IMC.
- Perform unit maintenance on assigned aircraft, armament, and avionics.
- Perform unit maintenance on assigned equipment (except medical and COMSEC equipment).

HEADQUARTERS AND HEADQUARTERS COMPANY

4-14. The HHC (figure 4-3) provides personnel and equipment for C2 functions of the ABTF, and security and defense for the CP. The HHC also provides unit level personnel service, UMT, logistics, and CBRN support.

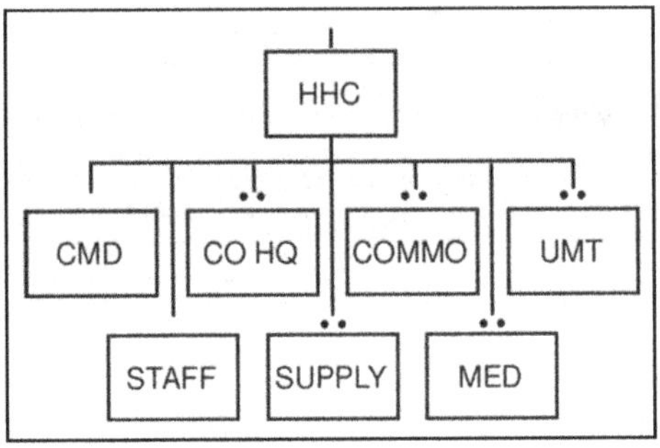

Figure 4-3. Headquarters and headquarters company

FORWARD SUPPORT COMPANY

4-15. The forward support company (FSC) (figure 4-4, page 4-5) has a company headquarters, field feeding section, distribution platoon, and ground maintenance platoon. The FSC provides field feeding, transportation, refueling, ammunition, and ground maintenance support, and coordinates with the ASB for additional support as required. Battalions contributing assets to the ABTF may also be required to contribute appropriate support assets to augment those in the ABTF or support those functions not normally organic to the battalion forming a TF.

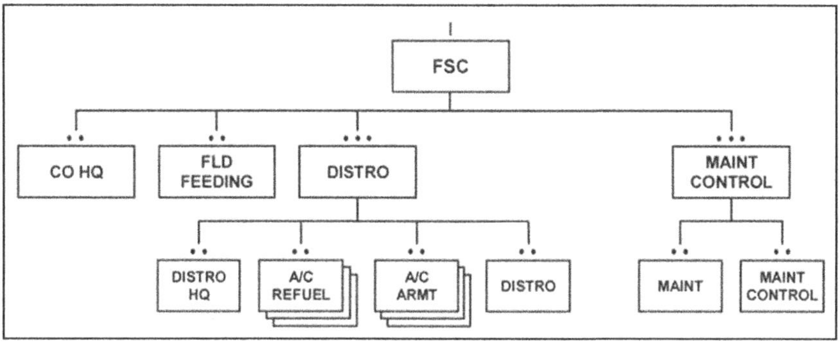

Figure 4-4. Forward support company

ATTACK RECONNAISSANCE ELEMENT

ORGANIZATION

4-16. The attack reconnaissance unit is comprised of one attack reconnaissance company (ARC) and one attack reconnaissance troop (ART).

Attack Reconnaissance Company

4-17. The ARC (figure 4-5) is organized with a company headquarters and two attack reconnaissance platoons with four AH-64D aircraft each.

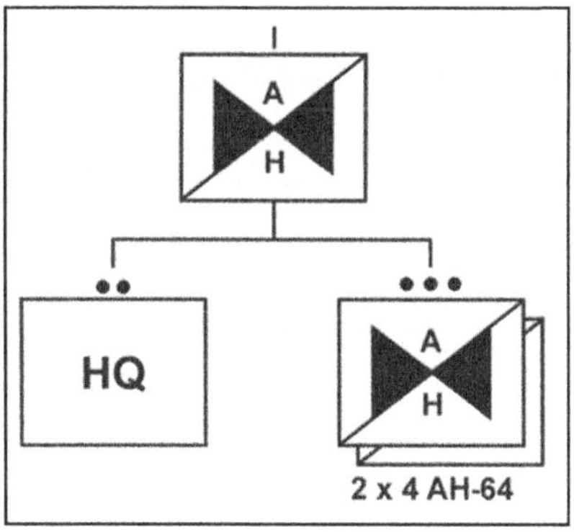

Figure 4-5. Attack reconnaissance company

Attack Reconnaissance Troop

4-18. The ART (figure 4-6, page 4-6) consists of a troop headquarters section and two attack reconnaissance platoons with five OH-58D aircraft each.

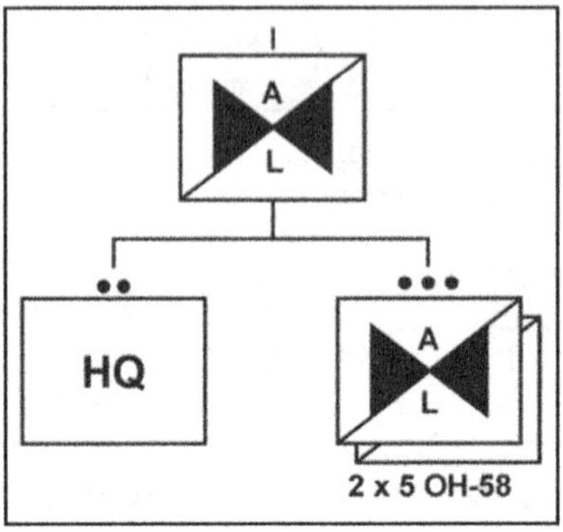

Figure 4-6. Attack reconnaissance troop

Mission

4-19. The primary missions of ARCs and ARTs are the following:

- Reconnaissance
- Security
- Movement to contact
- Attack

4-20. Specific tasks and roles of ARCs and ARTs include—

- Destroy enemy formations.
- Support friendly maneuver forces through CCA.
- Provide aerial escort and suppressive fires in support of air assault and other TACOPS.
- Provide mobile firepower to exploit the effects of artillery and other indirect fire.
- Conduct screening operations maintaining surveillance and providing early warning of contact with enemy forces.
- Conduct guard and covering force operations as part of a larger force.
- Conduct raids.
- Conduct team operations in multiple locations simultaneously.
- Conduct CBRN reconnaissance including aerial radiological surveys.

ASSAULT HELICOPTER COMPANY

Organization

4-21. The assault helicopter company (AHC) is organized with a company headquarters section and two assault helicopter platoons of five UH-60 aircraft each. When organized as part of an ABTF, AHCs may have a section of two EUH-60 aircraft from a GSAB equipped with A2C2Ss. Figure 4-7, page 4-7, depicts an AHC with GSAB A2C2S augmentation.

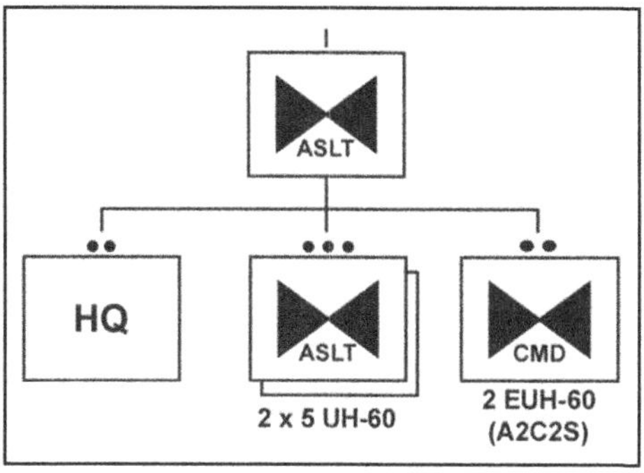

Figure 4-7. Assault helicopter company with general support aviation battalion Army airborne command and control system augmentation

MISSION

4-22. Primary AHC missions are the following:

- Air assault.
- Air movement.
- Command and control support.
- CASEVAC.
- Personal recovery support.

4-23. Specific AHC tasks and roles include—

- Conduct wet-hawk operations.
- Conduct SPIES, FRIES.
- Conduct team insertion/extraction.
- Conduct aerial mine delivery operations (Volcano-equipped UH-60).
- Conduct psychological operations missions (leaflet drop, speaker missions).
- Evacuate downed aircraft and personnel, when required by METT-TC or the nonavailability of logistic support aircraft.

HEAVY HELICOPTER PLATOON

ORGANIZATION

4-24. The cargo helicopter platoon (figure 4-8, page 4-8) consists of a platoon headquarters section and one section with four CH-47 aircraft.

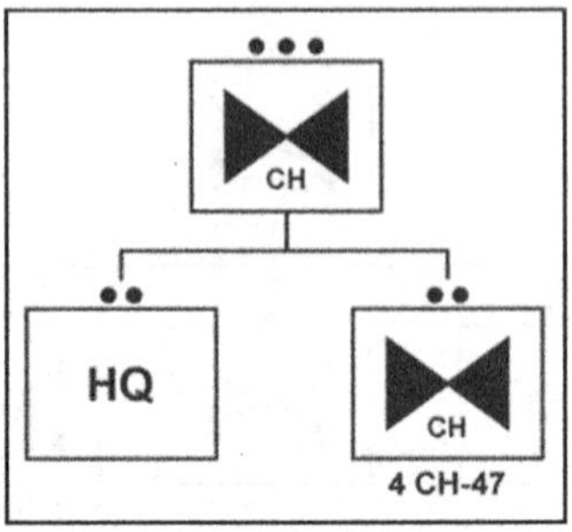

Figure 4-8. Cargo helicopter platoon

MISSION

4-25. The primary missions of the cargo helicopter platoon is the following:

- Air assault.
- Air Movement.
- CASEVAC.
- Personal recovery support.

4-26. Specific cargo helicopter tasks/roles include—

- Artillery raid.
- Team insertion/extraction.
- Fat Cow refueling operations.

4-27. Because of CH-47 characteristics, cargo helicopters can perform high-altitude operations and oversized, heavy, and special munitions movement.

FORWARD SUPPORT MEDICAL EVACUATION TEAM

ORGANIZATION

4-28. The forward support medical evacuation team (FSMT) (figure 4-9) consists of three HH-60 aircraft supporting 24-hour operations.

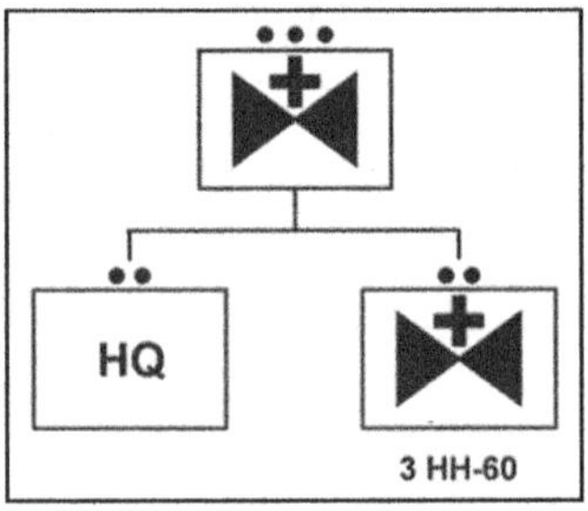

Figure 4-9. Forward support medical evacuation team

MISSION

4-29. The primary mission of the FSMT is aeromedical evacuation; however, it also provides the following:

- Patient movement between MTFs (patient transfers).
- Class VIII resupply.
- Joint blood program support.
- Medical C2.
- Movement of medical personnel and equipment.
- Air crash rescue support.

AVIATION MAINTENANCE COMPANY

ORGANIZATION

4-30. The aviation maintenance company (figure 4-10) must be structured and augmented to provide aviation unit-level field maintenance to all assigned or attached aircraft. Battalions contributing helicopters to the ABTF must also supply appropriate maintenance assets to support those aircraft not normally organic to the ABTF.

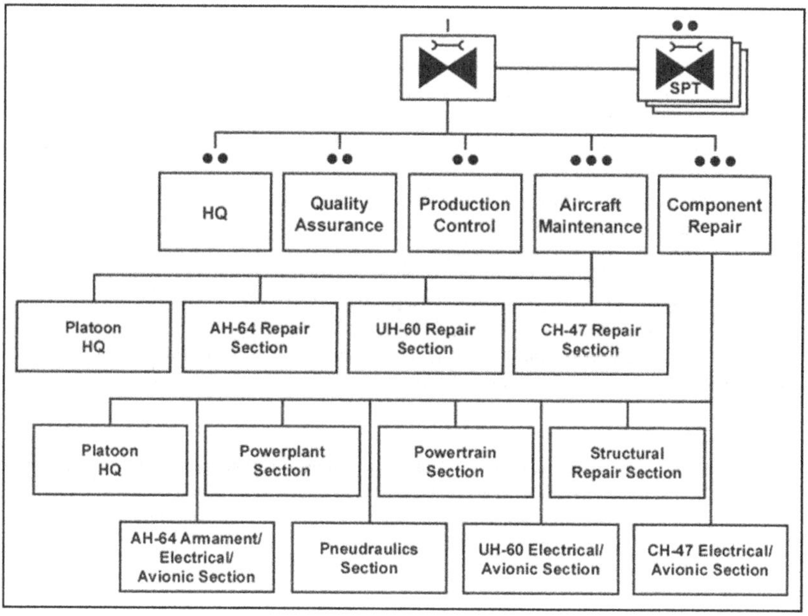

Figure 4-10. Sample aviation maintenance company

MISSION

4-31. The aviation maintenance company is designed to provide DS; tailored, critical, and routine aviation field level maintenance; battle damage assessment and repair (BDAR); as well as maneuver sustainment to its supported ABTF. The aviation maintenance company has the ability to perform the following functions:

- Repair aircraft, components, avionics, and armament.
- Provide aircraft unit-level maintenance, combat maintenance, and BDAR.
- Provide armament support collocated with FARPs.

SECTION III – IN-THEATER OPERATIONAL CONSIDERATIONS

4-32. An ABTF can be formed after units have deployed and are established in-theater. The ABTF may be established for a specific period of time or operation. If an ABTF is formed in the middle of a campaign, additional operational factors must be considered. These considerations include support relationships, unit status, operational updates, and aircraft specific considerations.

ACTIONS UPON NOTIFICATION

4-33. Upon notification that an additional aviation unit will be under the control of the ABTF, the main CP requires the following information from that unit:

- Current chain of command.
- An officer to assist in operational planning and ensure proper mission employment.
- Number of aircraft to be attached, OPCON, or TACON.
- Call signs and frequencies of elements.
- TACSOP of incoming element.
- Current operations being conducted, if any.

4-34. Initial coordination should provide the following additional information:

- Current location.
- Location in the next 24 to 48 hours.
- Attachments and detachments.
- Mission configured aircraft (such as Fat Hawk/Fat Cow).
- Available aircraft's number, type, and model.
- Planning allowable cabin load (ACL) per aircraft, if different due to environmental factors.
- Number of day and night crews available.
- Experience level of crews.
- Logistics requirements.

4-35. The ABTF S-4 requires the following information to ensure adequate support for the new unit:

- Current aircraft availability and bank time status.
- Current mission equipment status (such as M-240 machine guns, forward area refueling equipment [FARE], internal fuel tanks).
- Expected aircraft and major system status in the next 24 to 48 hours.
- Current fuel and ammunition status.
- Parts status and parts location.
- Expected daily consumption rate for fuel and other consumables per aircraft flight hour.
- The portion of organic maintenance elements accompanying them.
- Any additional personnel accompanying the element, such as cooks, operations, and class III personnel.

- For long term operations, the phase maintenance or progressive phase maintenance interval for the aircraft.
- Location of current aviation intermediate field maintenance support unit.

SPECIAL CONSIDERATIONS

4-36. Each unit has unique challenges. Specific things to consider include—

- Whether the element is to be collocated within the ABTF AA.
- The unit's ability to secure its portion of the perimeter.
- Aircraft parking and separation requirements (substantially greater for CH-47s than with other aircraft).
- Hot refuel requirements (CH-47s require a minimum of 30 minutes at refuel point).
- Other factors such as rotor wash from hovering CH-47s pose a significant risk to other aircraft and temporary structures without proper separation.

OPERATIONAL REQUIREMENTS

4-37. Units attached or placed OPCON or TACON to also require up-to-date operational information. Often elements coming from higher levels do not have access to tactical information at division level and below. At a minimum, information provided by the ABTF commander and staff to the incoming element should include the following:

- Operational reference material.
- Current operations.
- AO general information.
- Upcoming missions.

OPERATIONAL REFERENCE MATERIAL

4-38. Current operational documents and products essential for smooth integration include—

- ABTF TACSOP.
- Aviation procedures guide.
- Helicopter landing sites.
- Kneeboard products.
- ROE.
- SPINS.
- ACOs and ATOs.
- Current imagery, if applicable.
- Communications and security information for the ABTF.
- Intelligence.

4-39. By-products of the current ABTF IPB necessary to gain SA include—

- Threat situation.
- Known threat locations.
- Threat actions in the vicinity of the aviation unit's AA (past 24 to 48 hours).
- Expected threat contacts in the AO.
- Probable COAs.
- Weather forecasts for anticipated mission times.
- Weather constraints affecting the mission.

Current Operations

4-40. Current operational information should include the following:

- ABTF main CP and TAC CP locations (current and future).
- Subordinate unit locations.
- FARP locations (current and future).
- Coordinated airspace for operating in the AO.
- AC2 measures in the AO.
- Operational coordination (from the ABTF).

4-41. A clear understanding in regard to current operations must be achieved between all elements of the ABTF. Basics requirements include—

- Readiness condition (REDCON) status.
- Launch decision points.
- Planning time lines.
- Overall current mission (two levels up).
- Command relationships.
- Initial planning graphics.
- Call signs and frequencies.
- Time hack.

Area of Operations General Information

4-42. General information on the AO should include—

- EW support.
- Decontamination sites.
- Medical support.
- Boundaries.

Upcoming Missions

4-43. If missions are being prepared for execution with ground elements, incoming elements require additional detailed information on their planned roles. This information should include—

- Ground commander's mission and concept of the operation.
- Ground commander's intent for aviation use.
- OPORD and graphics.
- Special equipment requirements.
- Locations of friendly TAC CPs, brigade support areas (BSAs), and battalion/TF combat trains.
- Front line trace of friendly units.
- Target lists from the FSE.
- Location of friendly artillery current and preplanned FPs.
- Location of friendly AD assets.
- Frequencies and call signs.
- Challenge and password.

SECTION IV – TASK ORGANIZATION CONSIDERATIONS

SYNCHRONIZATION OF ASSETS

4-44. TF elements often come from different organizations and locations. These organizations may be active Army or RCs. Locations may be in the continental United States (CONUS) or outside the continental United States (OCONUS). TF elements have varying degrees of mission proficiency and may have different procedures for tactical and administrative tasks. Thus, the ABTF commander and staff, and subordinate commanders and leaders must synchronize TF assets.

STANDARDIZATION OF PROCEDURES

4-45. The immediate challenge facing commanders and their staffs is the requirement to organize and train the ABTF as quickly and thoroughly as possible. Because the ABTF must function as a unit or be further task organized, it is essential to operate from a common SOP. SOPs are tailored to the unit's METL, theater procedures, and expected OEs.

CHARACTERISTICS

4-46. The ABTF is not a standing unit with a published TO&E; its organizational structure must be tailored to meet mission requirements placed upon it.

Structure

4-47. The ABTF should be structured to provide—

- A balanced mix of attack reconnaissance and assault company-sized units and appropriate portions of the GSAB providing the ABTF commander with a full range of aviation capabilities.
- The capability to create aviation teams of mixed aircraft at platoon and company level capitalizing on the synergy inherent in a mix of attack reconnaissance, assault, and GS lift helicopters.
- C2 support to ABTF's higher headquarters (two additional EUH-60s in the assault company).
- ATS to meet the ABTF's own requirements.
- An aeromedical evacuation capability. If multiple ABTFs are deployed in-theater, only one ABTF may require aeromedical evacuation aircraft as part of the TF. To preclude complications involved in separate organizations (parts, critical skills, airspace management) not normally reporting to each other, the command relationship between the ABTF and aeromedical evacuation element is critical. Attachment is recommended.
- The capability to accept further aviation attachments in any form (U.S. or allied attack reconnaissance, air assault, heavy lift, FW, or UAS).
- The capability to accept attachment, OPCON, or TACON of ground units.
- An augmented aviation maintenance company capable of supporting all aircraft in the ABTF. In most cases, allied aircraft and UAS will come with their own maintenance structure; however, all aspects of support must be determined before the maintenance structure can be completed.
- Integrated aviation intermediate field maintenance support.
- Management of aviation maintenance requirements exceeding its organic capabilities. It is often necessary to reachback to the major command supporting the overall operation, and Army organizations and vendors in the U.S. This reachback capability is necessary when ensuring critical parts, specialized skills, and timely support are available if required.

Responsive

4-48. The ABTF is a fast-moving force providing versatile means of attaining accurate SA, while seamlessly linking with other Army and joint systems. Additionally, when assigned AH-64 helicopters, the

ABTF has the capability to disrupt and potentially destroy a regimental-sized armored force. The ABTF can conduct the following operations while mitigating risk in rapidly changing situations.

- The ARC expands the commander's area of influence throughout the OE utilizing armed reconnaissance, attack, and coordination of indirect fires.
- Air movement of forces and delivery of aerial mines with its AHC help overcome the effects of complex terrain while seizing the initiative.
- Cross-attached organizations consisting of attack reconnaissance, assault, and GS helicopters supported by UAS provide the ability to conduct operations from MCO through civil support operations worldwide and within the U.S.
- Air assets can operate from naval ships, greatly extending the joint commander's area of influence and employment options in coastal areas.
- UH-60 and AH-64 aircraft may be able to self-deploy. Assigned CH-47 and C-12 aircraft can self-deploy.
- The ABTF has the ability to take full advantage of en route planning and rehearsal systems, as these systems become available.
- The ABTF can prepare to fight soon after arrival with minimum reliance on RSOI assets within the AO.

Deployable

4-49. The ABTF is strategically deployable and can be task organized internally and externally to meet unique OEs. Its hybrid capabilities of attack reconnaissance and assault permit a full range of integrated aviation operations immediately upon arrival in-theater. An airlifted ABTF may be a suitable economy of force asset in support of the JTF until other aviation brigade assets can deploy by sea. The AH-64 and UH-60 can deploy by C-5 and C-17 aircraft. The OH-58D is transportable aboard C-130 and C-141 aircraft, as well as larger strategic air lifters.

Agile

4-50. The ABTF design, organization, and inherent flexibility to rapidly attach, detach, and cross attach assets ensure a TF with the physical agility necessary for commanders to maximize the operational potential of the force.

Versatile

4-51. Combining attack reconnaissance and assault aircraft in the same TF provides the ABTF versatility normally associated with an aviation brigade, even though the TF may have only a third of the aircraft found in the brigade. Additionally, the ABTF's modularity allows task organization between multiple ABTFs to quickly mass specific aircraft system capabilities when multiple ABTFs are deployed.

Lethal

4-52. The ABTF can provide fires with its ARCs/ARTs while rapidly refueling and rearming these assets with its AHC and cargo helicopter assets. This force, in conjunction with ground combat forces and augmented with joint and combined forces, becomes an even more lethal combat multiplier.

Survivable

4-53. The mobility of the ABTF greatly enhances its survivability. The ability to stand off from both enemy direct fire and AD weapons, coupled with its array of ASE, further reduces its vulnerability.

Sustainable

4-54. The ABTF's motor maintenance, mess, medical, supply elements, and aviation maintenance company (when properly structured and supplemented by appropriate ASB sustainment and maintenance) allow it to sustain itself while operating independently, under aviation brigade control, or attached to ground maneuver units.

TRAINING

4-55. The ABTF must train as a TF together with its supported BCT before deployment. Such training allows proficiency in the most critical missions. The ABTF must refine, internalize, and practice a common SOP. Critical to preparation is a full understanding of in-theater requirements.

SECTION V – UNIT CONSIDERATIONS

BATTALION AND ABOVE

4-56. The parent brigade of the ABTF headquarters, remnants of the core battalion that formed the ABTF, and the battalions contributing units assigned to the ABTF must adjust to the creation of the ABTF. All parties must coordinate in formation of the ABTF and adjust operations to compensate for reduced capabilities and resources.

PARENT BRIGADE

4-57. The aviation brigade commander ensures the ABTF is properly resourced and trained to operate in accordance with the mission of the ABTF. He or she then assesses the capability of remaining battalions to maintain and continue normal brigade operations. The aviation brigade commander and his or her staff must adjust to compensate for the altered force structure.

4-58. The brigade determines the status of those stay behind forces of the core TF battalion. Based on remaining personnel and equipment, the unit may continue normal operations with a reduced staff and interim commander. If remaining resources preclude normal operations, the brigade commander may attach the remnants to another battalion within the brigade (the core battalion of an ABTF is an AHB receiving units and elements from the AHB, GSAB, and ASB). It is very likely one of the AHB's companies will remain behind to support brigade missions. Often the brigade is not only concerned with the formation, training and deployment of the ABTF; it must also consider employment and support of the remaining AHC.

4-59. The brigade provides support to the remnants of the core battalion that formed the ABTF. If not reassigned to another fully capable battalion, remaining personnel must continue to have support, or be given permission to access support, from other brigade or installation assets.

STAY BEHIND FORCES

4-60. ABTF organizations may not include all core battalion organic resources, a portion of the core battalion will continue to operate and provide support from an AO or home station.

4-61. The ABTF headquarters must first determine battalion personnel and equipment that will fall under ABTF control. The remaining elements must then reorganize into a battalion-minus structure to continue operations and prepare to support or reinforce the ABTF as necessary.

4-62. In close coordination with the parent brigade, the ABTF commander must give the rear detachment commander training and operational guidance for the stay behind forces. This includes accommodations to compensate for those capabilities under ABTF control not accessible to the rear detachment.

CONTRIBUTING BATTALIONS

4-63. The battalions and companies who contribute forces to the ABTF must also adjust to operating with diminished forces, each of which is left with varied degrees of capability. The commander must adjust the unit's operations compensating for the altered force structure.

4-64. If the ABTF headquarters leaves a portion of its organization in the rear and those elements are not capable of conducting normal operations, the aviation brigade commander may attach the remnants to one of the contributing battalions. The contributing battalion may then be faced with supporting two completely separate missions, and may be required to carry on normal operations at an AO or home station while supervising the remnants of the ABTF.

COMPANY AND BELOW

FLIGHT COMPANY

4-65. Flight companies are dependent upon the aviation maintenance company, HHC, FSC, and the ABTF staff for operations, intelligence, personnel service, maintenance, and logistics support. Accordingly, they have little or no capability for independent operations. They are dependent on—

- The aviation maintenance company for maintenance above that provided by its assigned crew chiefs and for class IX (air) supply support.
- The HHC and FSC for religious, paralegal, unit-level combat health support, personnel and administrative services, and unit-level logistics support, including supply (except class IX air), automotive and communications maintenance, food service, CBRN support, and unit-level supply support for classes III and V.
- The ABTF S-2 for intelligence.

AIRCRAFT OPERATIONS

4-66. Operational limitations for ABTF aircraft and units are similar to those when assigned to their parent units. Examples of operational limitations include—

- Extreme environmental effects (temperature, altitude) reduce payloads and flight endurance.
- If used, auxiliary fuel tanks or Kevlar blankets limit allowable cargo load (ammunition loads, personnel, and equipment).
- Use of seats in assault aircraft limits allowable cargo load but increase troop protection during a crash sequence.
- Weather, in some cases, may preclude aviation operations (visibility, ice, high winds, and excessive turbulence).
- Limited visibility from weather effects (fog, heavy rain, blowing snow) or battlefield obscuration (smoke, dust) may limit sensor and optic capabilities, observation, target acquisition (TA), engagement ranges, and speed of aircraft movement.
- Low ceilings affect target engagement options.
- Weather may require aviation deployments under IMC. OH-58 and AH-64 aircraft are not IMC rated.
- Units have limited capability to secure unit AAs and concurrently conduct required operations and aircraft maintenance. Mutual support can reduce the amount of dedicated security needed by the ABTF.
- Crew endurance and aircraft maintenance requirements may impact aircraft availability.
- Terrain may limit the availability of adequate AAs, PZs, or LZs.

ATTACK RECONNAISSANCE COMPANY/TROOP

4-67. An ARC consists of eight AH-64 aircraft and an ART consists of ten OH-58D aircraft.

4-68. The AH-64 can be deployed by C-5 and C-17 aircraft. The OH-58D is transportable aboard C-130 and C-141 aircraft, as well as C-5 and C-17 aircraft.

ASSAULT HELICOPTER COMPANY

4-69. The AHC has ten aircraft and is usually organized into two platoons of five aircraft each. The AHC is often augmented with a section of two A2C2S-equipped EUH-60 aircraft from the GSAB.

4-70. The AHC must continually balance lift requirements with C2 requirements. The commander organizes the aircraft available for assault operations to meet mission requirements.

4-71. The AHC's UH-60 can be deployed by C-5 and C-17 aircraft.

COMMAND AVIATION COMPANY

4-72. The CAC has two flight platoons; one C2 platoon with four EUH-60 aircraft, and one GS platoon with four UH-60 aircraft. One of these UH-60s is fitted with the A2C2S A mission kit.

4-73. CAC aircraft are often further placed OPCON to the maneuver commander. When utilizing a A2C2S aircraft, the commander enjoys maximum mobility without sacrificing access to information or jeopardizing continuity of operations due to CP relocation.

4-74. The CAC's EUH-60s and UH-60s can be deployed by C-5 and C-17 aircraft.

HEAVY HELICOPTER COMPANY

4-75. Heavy HCs are organic to all aviation brigades. Each GSAB has one heavy HC consisting of a company headquarters, and three flight platoons with four aircraft each. The ABTF can receive one or more of these heavy HC platoons.

4-76. Because many of the theater aviation brigade's heavy HCs are RC, collective training prior to employment is essential for maximum operational capability. Units may activate and deploy as part of a rotation of forces supporting ABTFs operating in stability operations.

4-77. Theater cargo helicopter assets remain under theater aviation brigade C2, even when dispersed in support of other organizations. Centralized control by theater aviation brigades ensures tasking and planning access and better focuses DS and GS assets on the higher commander's priorities. However, they may operate under control of the ABTF.

AIR AMBULANCE MEDICAL COMPANY

4-78. The AAMC consists of a company headquarters and four FSMTs. Each FSMT consists of three HH-60 aircraft that support 24-hour operations. Each aircraft can be individually or group deployed in support of the ABTF.

4-79. Whether deployed as a unit or an individual FSMT, AAMC assets are dependent upon the supported unit for food, fuel, security, intelligence, sector communications, wheeled vehicle maintenance, and all classes of supply except class IX (air).

4-80. Air ambulance assets can self-deploy if equipped with extended range fuel systems (ERFSs). The AAMC's HH-60s can also be deployed by C-5 and C-17 aircraft.

SECTION VI – EMPLOYMENT PRINCIPLES

AVIATION BATTALION TASK FORCE

4-81. The ABTF's primary role is to plan and execute the entire spectrum of aviation operations and can expect to operate anywhere in the assigned unit's AO. The ABTF allocates resources based on METT–TC, scheme of maneuver, available assets, and higher headquarters commander's priorities.

4-82. Recent conflicts in Kosovo, Afghanistan, and Iraq illustrate a noncontiguous battlefield is fast becoming the normal. As such, it leaves ABTF aircraft increasingly vulnerable to small arms and rocket-propelled grenade fire anywhere on the battlefield during conduct of air movement. Security provided by attack reconnaissance aircraft may be essential for single- and dual-aircraft missions based on mission analysis and theater SOP.

4-83. In support of stability operations, the ABTF may transport allied leaders and perform team insertion/extraction. Aircraft may operate from remote base camps supporting patrolling forces and reaction teams, as well as counter-drug efforts.

ATTACK RECONNAISSANCE COMPANY/TROOP

ATTACK

4-84. During MCO, ARCs/ARTs are most effective when used en mass and providing continuous operations on enemy flanks and rear. During stability operations, ARCs/ARTs are most effective when used in teams. The ABTF may be tasked to conduct attack operations as a TF. Augmentation permits a balanced ABTF sustaining continuous attack operations. Although the ABTF has a limited number of attack reconnaissance aircraft, this limitation is somewhat offset by its own lift assets used for logistics support. The AHC provides substantial flexibility in resupply of class III/V and insertion of ground troops in blocking positions or OPs. The AHC operates under TF control in GS of the ARC/ART during attack operations. Based on METT-TC, the ABTF staff must backward plan from operations in the EA to preparatory actions in the AA. Planning may include method of employment (continuous, phased, or maximum destruction), occupation of BPs, holding areas, and air movement routes (include passage of lines, if required).

RECONNAISSANCE AND SECURITY

4-85. The ABTF is capable of conducting all security missions including screen, guard, and cover missions; however, it normally participates in guard and covering force operations as part of a larger force. To act as the covering force headquarters, the ABTF requires additional ground troops and DS artillery.

4-86. The firepower, sensors, and maneuverability of the ARCs and ARTs coupled with the logistics and assault flexibility afforded by the AHC, provide the ABTF with significant advantages when conducting traditional air cavalry missions.

4-87. When augmented with ground forces and UAS, the ABTF operates as a reaction force developing the situation, occupying ground OPs, seizing key terrain, and conducting raids. METT-TC determines whether the ABTF commander operates with companies pure or task organized.

ASSAULT HELICOPTER COMPANY

4-88. Assault helicopters operate throughout the battlefield day and night as a fully integrated member of the combined arms team. An AHC in an ABTF can expect to conduct maneuver, support, and sustainment operations. Even though the UH-60 lacks sophisticated weapons and sensors, when pressed, the ABTF may use the AHC to conduct limited reconnaissance and surveillance in accordance with METT-TC.

AIR ASSAULT AND AIR MOVEMENT

4-89. The ABTF can perform company-sized air assaults or air movements, and may, when augmented, conduct larger scale operations. TF air assault operations are normally conducted with the ARC/ART providing overwatch fires, route reconnaissance, and security. Successful operation execution is based on a careful analysis of METT-TC factors and a detailed, precise, reverse planning sequence.

4-90. The supported unit must supply all slings and rigging equipment for air movement and air assault operations. The supported unit is also responsible for preparing all loads for movement. Failure to establish this responsibility early in the planning process leads to major mission delays and possible mission failure. During training and mission planning it is wise for aviation leaders to assist ground units in training proper load and rigging operations.

CASUALTY EVACUATION

4-91. For casualty rates exceeding the capabilities of aeromedical evacuation elements, assault and cargo helicopters may be employed for CASEVAC. UH-60 and CH-47 aircraft can be employed using several different configurations for CASEVAC operations. The number of casualties transported is dependent on the type of casualty (ambulatory versus litter) and severity of injuries and wounds.

COMMAND AVIATION COMPANY

4-92. The CAC provides a means by which air and ground commanders can rapidly traverse and see the battlefield. The CAC also provides airborne C2, aerial retransmission, and GS, as directed.

4-93. The CAC supports BCOTM requirements with its platoon of A2C2S-equipped aircraft employed in support of the higher headquarters' command group, and maneuver and aviation brigade commander's C2 requirements. A2C2S gives the commander an enhanced capability to C2 assets over extended distances. With networked digitized communication systems, commanders and staffs can assimilate significantly greater amounts of data faster and with greater clarity. The A2C2S primary roles are—

- BCOTM.
- Ground and aerial TAC CP.
- Early entry CP.
- First responder during national disasters.

4-94. The onboard communications linkages allow for continuous contact between the commander and committed forces. These linkages also help maintain SA, issue and receive FRAGOs with graphics, synchronize fires and maneuver, and extend coverage. Refer to appendix B for additional information on A2C2S.

4-95. The GS platoon provides GS as directed. These aircraft are also used in support of aircraft sustainment requirements. CAC aircraft allow commanders to maintain communications with their forces and provide timely information supporting critical decisions without sacrificing mobility and efficiency.

HEAVY HELICOPTER COMPANY

4-96. Heavy helicopter units can conduct support and sustainment operations day and night throughout the OE. They routinely transport heavier equipment such as high mobility multi-purpose wheeled vehicles (HMMWVs) and howitzers, and support logistics efforts. The ABTF commander, based on the ground commander's support plan, determines how to best employ cargo helicopters. METT-TC influence the missions assigned to cargo helicopter units.

4-97. HvyHC elements may insert and extract long range surveillance detachment Soldiers, and transport other intelligence equipment, such as Prophet Systems (signals intelligence/EW) and HMMWVs in support of AATFs and attack reconnaissance helicopter raids. The heavy HC requires detailed intelligence of en route threats during these missions, as well as potential threats and terrain/weather considerations at the LZ. A similar level of intelligence is necessary for false insertions that may be part of a team insertion or separate diversionary mission.

4-98. Heavy helicopter elements provide the ABTF with tremendous capability. The CH-47D can transport up to 31 combat loaded troops. Air assault missions for cargo helicopters may include artillery raids, externally transporting M105 (105-millimeter) or M198 (155-millimeter) howitzers, their prime movers, and ammunition.

4-99. Air assault forces normally arrive in an LZ with only minimum essential supplies. Heavy helicopters must provide follow-up support to an air assault force by providing resupply (normally classes I and V). Heavy helicopters must be prepared to rapidly refuel and return to a designated PZ to load additional critical supplies or wait for a designated serial launch time. Air assault staff planning must include resupply of air assault forces as part of the overall operation.

4-100. Heavy helicopter units are subject to operating limitations the commander must consider when planning for cargo helicopter employment. These limitations include—

- Availability of adequate PZs and LZs due to terrain.
- A larger IR signature making cargo helicopters more vulnerable to IR missiles.
- Extensive fuel, maintenance, and parts support required for extended operations.

Emergency Resupply

4-101. Units conduct emergency resupply from logistics support areas forward to a unit in contact or a unit having recently broken contact. This critical rapid resupply operation is often necessary for ground forces to continue the attack. Consideration must be given to the friendly and enemy situation, AC2 measures established, friendly weapons control status, and call signs and frequencies of the receiving unit before the mission can be executed. Time is essential during emergency class V resupply. Prior coordination by the ABTF staff can result in anticipation of this mission and ultimately a quicker response time.

Aviation Resupply

4-102. Heavy helicopters may also be used to move classes III and V supplies forward to establish jump FARPs in support of ongoing aviation operations. CH-47s may be given missions to transport helicopter ammunition forward from main FARPs or ammunition transfer points (ATPs) to sustain the rapid tempo of attack reconnaissance helicopter operations.

Fat Cow

4-103. The CH-47, equipped with improved ERFS located in the heavy bay, can operate up to four refuel points. The system can be equipped with two 800-gallon fuel cells. The Fat Cow site configuration is depicted in figure 4-11.

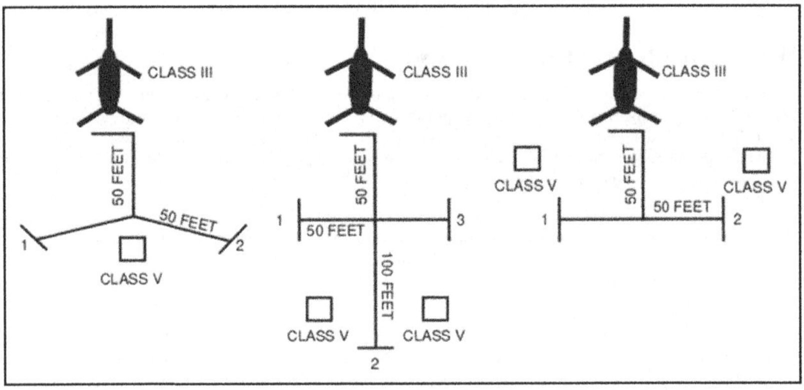

Figure 4-11. CH-47 Fat Cow forward arming and refueling point site

Aircraft Recovery

4-104. Combat operations result in a greater demand for operational aircraft and a large increase in flying hours. These increased requirements are further complicated by higher attrition and battle damage, which create shortages of repair parts and replacement aircraft. To offset these shortages and maintain an effective combat aviation force, the rapid recovery and repair of aircraft is essential. The CH-47 is the only helicopter able to recover every aircraft in the U.S. Army inventory, including itself.

CASUALTY EVACUATION

4-105. If the casualty rate exceeds the capabilities of aeromedical evacuation elements, cargo helicopters may be employed for CASEVAC. CH-47s can be employed using several different configurations for CASEVAC operations.

- **Seats folded**. With seats folded up, the number of casualties transported is dependent upon the type of casualty (ambulatory versus litter) and severity of injuries and wounds.
- **Seats down**. With seats folded down, the lifting capacity for patients is reduced to 30-seated ambulatory casualties transported with an accompanying medic, while litter patients can be placed on the floor as directed by the aircrew.
- **Litter configuration**. CH-47s can be equipped with a litter kit giving them the capacity to transport 24 litter patients. In this litter configuration, the CH-47 seats are replaced with six tiers of litters, four litters high. The CH-47 litter support kit consists of poles and supports only. Medical assets supporting the unit must provide litters and tie-down straps.

AIR AMBULANCE MEDICAL COMPANY

4-106. Army AAMCs are the only dedicated aeromedical assets within the Department of Defense (DOD). In both the DS and GS roles, FSMT efforts should not be rigidly tied to any area or supported unit. FSMTs must retain flexibility in employment.

4-107. Air ambulance assets are a limited resource and are located where they are most needed. This can be with troops most often engaged in combat, high population density areas, area of famine or disease with high civilian casualties, refugee areas, or geographically centralized locations.

4-108. FSMTs performing area or GS accomplish the patient transfer mission that develops between FSMTs and MTFs, between MTFs, and between MTFs and intertheater movement locations. These FSMTs must also be responsible for the DS mission within their immediate vicinity.

4-109. Air ambulance headquarters may collocate with the supported BCT, AHC or TF CP as required. Task organization of aeromedical evacuation assets requires careful planning to ensure adequate resources (maintenance, flight operations and communications capabilities) are provided by the ABTF headquarters or appropriate supported element. Refer to FM 1-113 for additional information on air ambulance employment.

Chapter 5
Logistics Operations

This chapter describes sustainment (maintenance and logistics) doctrine with a focus on the ASB, and how the battalion optimizes organizational effectiveness while balancing sustainability, mobility, and survivability against requirements for rapid strategic deployability. It also provides a general explanation of TTPs that can be employed to exploit the ASB's range of logistics capabilities and ensure versatility across the full range of potential requirements.

SECTION I – INTRODUCTION

Contents

5-1. Today's OE significantly impacts Army logistics. In a theater of operations with combat forces widely distributed and operating in noncontiguous areas, support must be provided in innovative ways, leveraging new technologies and ideas.

5-2. Current OEs place mid-grade and junior leaders in complex situations, with the potential of having international, informational, and political ramifications. It is in these complex situations that leaders must recognize and solve challenging tactical action issues.

OPERATIONS

5-3. The aviation brigade receives sustainment from various elements depending on the logistics organizational structure at the brigade and division sustainment brigade. Brigade and battalion XOs are responsible to their respective commanders for overwatching sustainment operations and inserting themselves where appropriate to ensure success. S-4s identify logistics requirements for the maneuver plan and provide them to the FSC, ASC, or coordinate with the ASB commander as appropriate.

COMMAND RESPONSIBILITY

5-4. The aviation brigade commander ensures sustainment is provided not only for his organic and attached elements, but also for any elements OPCON to or supporting units. The brigade S-4 coordinates logistics for the attachments and verifies who provides sustainment and how to request support for attachments. When a unit is attached to the aviation brigade, the attachment should bring an appropriate modular unit of logistics assets.

5-5. These assets are controlled by the unit they support. They can be attached to the ASB, the aviation battalion's FSC, or the aviation maintenance company in accordance with attachment instructions. The attached unit leader must coordinate with the brigade S-1 and furnish him or her a copy of the unit battle roster as well as provide the status of all key elements of equipment to the brigade S-4. Thereafter, the attached unit submits reports and requests support according to the aviation brigade SOP.

PLANNING FOR LOGISTICS OPERATIONS

5-6. The brigade S-4, brigade surgeon, and ASB support operations officer (SPO), operating from their respective CPs, monitor sustainment operations and ensure appropriate collaboration and synchronization of support. They use the logistics estimate, a product of the logistics preparation of the battlefield (LPB), to determine logistics capabilities, anticipate support requirements, identify and resolve shortfalls, and develop support plans. In addition, they work with their respective S-2s to develop the enemy threat to logistics operations from the IPB. They integrate all planning to develop, collaborate, and synchronize logistics with maneuver and fire plans. Logistics planners must thoroughly understand the mission, tactical plans, and commander's intent and develop the LPB. This culminates during the MDMP with a fully developed and integrated logistics/force health protection plan. Table 5-1 provides each MDMP step's inputs, actions, and outputs. Use of C2 products by the logistics planners and ASB commander is very helpful for developing SU. Throughout this entire process, staff estimates are also used to assist with maintaining the commander's SU.

Table 5-1. Sustainment aspect of military decisionmaking process—inputs, actions, and outputs

MDMP Step		
Inputs	***Actions***	***Outputs***
Receipt of mission and mission analysis		
Higher HQ WARNO or OPORD. Facts from higher, lower, & adjacent logistics planners. Higher HQ LPB & staff LPB products. Enemy COA from S-2. HVTs by phase or critical event. Facts from logistics assets. Cdr's initial logistics guidance. Staff estimates. Constraints & ROE.	Understand higher maneuver plan. Conduct logistics staff estimate - organize & analyze facts. Identify specified/implied tasks. Determine & portray friendly & threat INFOSYS capabilities & vulnerabilities. Translate status of logistics assets into capabilities/limitations. Analyze effects of LPB on sustainment. Develop draft desired logistics effects. Identify logistics related CCIR & essential elements of friendly information (EEFI). Identify logistics constraints/restrictions. Obtain CDR's initial logistics priorities.	Initial WARNO upon mission receipt. Logistics portion of mission analysis brief (end state analysis, logistics effects development). Draft logistics radar frequency interferometers. Recommend logistics tasks ROE guidance. Logistics CCIR/EEFI inputs. Initial logistics/force health protection rehearsal guidance. CDR approves initial logistics or modifies. CDR gives other sustainment guidance. WARNO after mission analysis brief.
COA development		
See outputs from previous step.	Determine logistics tasks for each COA. Allocate logistics assets to sustain. Allocate logistics assets forces to each IO task. Identify requirements for additional resources. Integrate sustainment triggers with maneuver COA. Analyze relative logistics combat power. Use battle calculus. Assist S-2 in ISR plan development to support logistics.	For each COA developed: • Concept of support • ISR Plan • Logistics affects • IO execution timelines as they pertain to logistics. • Input to force protection plan • Refined logistics tasks.

Table 5-1. Sustainment aspect of military decisionmaking process—inputs, actions, and outputs

MDMP Step		
Inputs	***Actions***	***Outputs***
	Prepare logistics portion of COA/sketch.	
COA analysis & COA comparison		
See outputs from previous step.	Wargame the brigade COA & integrated logistics plans vs. enemy COAs. ID coordination requirements to produce synchronization matrix. Synchronize logistics effects. Finalize logistics tasks. Modify/refine inputs as required. Refine & test logistics plans.	Final Drafts: Paragraph 4 Logistics annex.
COA approval & orders production staff supervision		
See outputs from previous step.	Approval briefing. Logistics plan briefed as part of each COA. Bde S-4 or ASB SPO presents logistics analysis.	Commander: Selects, modifies or approves COA. Bde S-3: Issue WARNO as required. Finalize logistics products. Issue logistics plan & annexes with OPORD. Logistics planner's backbrief. Manage refinement. Rehearsals.

Logistics digital planning tools

5-7. The LPB and MDMP are tools that enable commanders to see, understand, act, and finish decisively. Commanders and battle staffs at all levels must have a thorough knowledge of these processes .

5-8. The aviation brigade's organic ASB provides distribution-based, centralized logistics and is fully digitally enabled with battle command sustainment support system (BCS-3), FBCB2, and movement tracking system (MTS). These digital enablers assist in providing a logistics common operational picture (LCOP) with communications linkages to the STAMIS. These systems enable the ASB support operations section to gain and maintain oversight of logistics requirements. The increasing use of assured communications and improvements in digital information technology provide the logistics operator and unit S-4 with tools needed to tailor the logistics package (LOGPAC). Through near real-time information, the aviation brigade staff and ASB staff are able to make timely adjustments in their support requirements.

5-9. The ASB commander, supported by his SPO and in conjunction with the aviation brigade S-1/S-4 and surgeon, closely monitors the implementation of the logistics concept of support as outlined in the brigade OPORD's logistics annex. The ASB commander adjusts logistics operations or shifts resources within his unit to account for a change in METT-TC factors or to replace lost logistics capabilities. Recommendations to surge logistics assets from units within the brigade, but not subordinate to the ASB, or request higher echelon support are made to the brigade commander by the ASB commander and brigade logistics staff.

LOGISTICS PREPARATION OF THE BATTLEFIELD

5-10. LPB is the process of gathering data against pertinent battlefield components, analyzing their impact on logistics, and integrating them into tactical planning so that support actions are synchronized with maneuver.

5-11. LPB is a conscious effort to identify and assess those factors, which facilitate, inhibit, or deny support to combat forces. Just as IPB is important to the conduct of actual combat operations, LPB is equally important to sustaining the combat power of the force. Working together, leaders must synchronize support actions with maneuver in a unified plan making logistics a factor in the success of a mission rather than a cause of failure. In addition to METT-TC, LPB focuses on determining the status and impact of the specific components that make up tactical logistics. It assesses how time and space requirements and restrictions of the battlefield affect support.

5-12. The process requires tacticians to understand the data needed by logisticians for planning and providing timely, effective support. It requires logisticians to understand the mission, tactical plan, and battlefield's time and space implications for support.

5-13. It is a coordinated effort to prepare the battlefield logistically. The basic steps in systemizing the process are—

- Determine battlefield data pertinent to support actions.
- Determine sources from which raw data can be derived.
- Gather pertinent data.
- Analyze collected data elements and translate them into decision information by assessing their impact on the mission and competing COAs.
- Integrate decision information into tactical planning by incorporating it in logistics estimates and brigade or battalion (as appropriate) plans and orders.

5-14. LPB products include the following:

- A logistics estimate.
- A visualization of the pending battle and logistics activity required by phase of operation.
- Anticipated logistics challenges and shortfalls, and their solutions.
- How, when, and where to position logistics units to best support the tactical commander's plan.
- A synchronized tactical and logistics effort.

Logistics Estimate

5-15. A logistics estimate is an analysis of logistics factors affecting mission accomplishment. The key concerns of logistics planners are the status of supply classes III, IV, and V; and the operational status of critical generators of combat power such as infantry Soldiers, aircraft, tanks, BFTs, Stryker, and other units that provide combat power. Logistics estimates at the combat battalion level are often not written, though at the ASB and brigade written products include combat power charts, periodic updated briefings, or commander's updates. They are frequently formulated in terms that answer the following questions:

- What is the current and projected status of maintenance, supply, and transportation?
- How much of what supplies are needed to support the operation?
- How will it be transported to where it is needed?
- What external echelons above brigade (EAB) support is needed?
- Can the requirements be met using host nation or throughput from EAB or are other techniques such as aerial resupply necessary?
- What are the shortfalls and negative impacts?
- What COAs can be supported?

Reconstitution

5-16. Reconstitution is a set of actions the commander plans and implements to restore his unit to a desired level of combat readiness commensurate with mission requirements and availability of resources. Reconstitution is a total process. Its major elements are assessment, reorganization, and regeneration. Although not a logistics function, reconstitution is often logistics intensive, especially regeneration. Reconstitution decisions rest with the commander. The commander, with his staff's support, assesses unit effectiveness (refer to FM 100-9). He or she does not base his reconstitution decisions solely on facts, figures, and STATREPs from subordinate units. His assessment relies also, and probably more importantly, on other factors. These include—

- Knowledge of his Soldiers.
- Condition and effectiveness of subordinate commanders and leaders.
- Previous, current, and anticipated situations and missions.

5-17. Planners must be prepared for mass casualties, mass destruction of equipment, and destruction or loss of effectiveness of entire units. The aviation battalion or companies catastrophically depleted or rendered ineffective are returned to combat effectiveness through this mission staging operation (MSO). Reconstitution differs from sustaining operations and replenishment and sustainment operations in that it is undertaken only when a unit is at an unacceptable level of combat readiness. Replenishment and sustainment operations are routine actions to maintain combat readiness. Weapon system replacement operations can be part of replenishment and sustainment operations.

Assessment

5-18. Assessment measures the unit's capability to perform a mission. Subordinate unit commanders assess their units before, during, and after operations. If a commander determines his unit is no longer mission capable even after reorganization, he or she notifies the aviation brigade commander. The aviation brigade commander either changes the mission of the unit to match its degraded capability or removes it from combat. Commanders can reconstitute their units by reorganization or regeneration to bring their units up to the necessary readiness level for the next mission.

Reorganization

5-19. Reorganization is the action taken to shift resources within a degraded organization to increase its combat power. Measures taken include cross-leveling equipment and personnel, matching operational weapons systems with crews, or forming composite units. It can be conducted down to and including company level. Depending upon the type of reorganization, the unit's own assets or higher echelon resources may be used.

5-20. Immediate battlefield reorganization is the quick and often temporary restoration of organizations conducted during an operation; for example, reorganizing on the objective and implementing the established succession of command is a quick method not requiring an MSO to achieve the desired results.

5-21. Deliberate reorganization is a permanent restructuring of the unit. It is the type of reorganization considered during reconstitution planning. Deliberate reorganization is supported with higher echelon resources (such as maintenance and transportation), and additional replacements and other resources may be made available during a MSO. Deliberate reorganization must be approved by the parent-unit commander one echelon higher than that reorganized. For example, the aviation battalion commander cannot approve the deliberate reorganization of an attached company; however, the parent aviation battalion commander or aviation brigade commander can.

Regeneration

5-22. Regeneration is incremental or whole-unit rebuilding through large-scale replacement of personnel, equipment, and supplies; reestablishing or replacing essential C2; and conducting the necessary training for the rebuilt unit. Regeneration is used when the unit has become combat ineffective.

5-23. The unit must be removed from combat to be regenerated during an MSO. The division or corps is responsible for the regeneration of aviation battalions. Aviation brigade regeneration is a theater responsibility. To regenerate a unit, the appropriate command must balance priorities for supplies, equipment, or other logistics requirements to include medical, and task the appropriate support organizations for needed support.

5-24. Aviation brigade regeneration could occur with redeployment back to its home station or an equally suitable environment. This requirement places the brigade in a location to effectively receive requisite resourcing (personnel and equipment) and a stable environment to retrain. A new aviation brigade is deployed to assume its mission requirements in the AO.

SECTION II – LOGISTICS FUNDAMENTALS

5-25. It is essential for all leaders, not just logisticians, to understand the fundamentals of supporting military operations. By understanding how the logistician is trained, manned, and equipped for sustainment operations, the supported commander will know what to expect. The following paragraphs discuss logistics characteristics and methods of resupply.

LOGISTICS SUPPORT DURING COMBAT OPERATIONS

5-26. Sustainment operations are inseparable from decisive and shaping operations, although they are not by themselves decisive or shaping. Sustainment operations occur throughout the AO, not just within the support areas. Sustaining operations determine how fast forces reconstitute and how far they can exploit success. At the tactical level, sustaining operations underwrite the tempo of the overall operation; they assure the ability of the brigade to take immediate advantage of any opportunity.

5-27. To support sustainment operations, logistics units must be able to conduct combat operations themselves. The enemy will use many different tactics to degrade the logistics infrastructure critical to support military operations. Unfortunately, the Army does not always know which tactics will by chosen by the enemy, so the aviation brigade's logistics assets must be prepared to defeat or destroy the enemy to mitigate enemy intent and action.

5-28. Aviation logistics units are equipped and manned to operate in a hostile environment while accomplishing sustainment operations. All commanders must acknowledge the basic concept that as security requirements increase, the ability to conduct sustainment operations decrease.

5-29. The aviation logistics commander must consider what level of force protection his unit can accomplish while still performing sustainment and support operations; for example, destroy Level I, defeat Level II with assistance, and employment of a TCF for Level III. This does not presume a 100 percent level of sustainment operations can occur 100 percent of the time. Sustainment may fluctuate depending on threat level and enemy operations. If the enemy threat is stronger than the ability of the aviation logistics unit to destroy or defeat, then the commander knows additional forces are required to sustain logistics operations at the level desired or risk their destruction.

5-30. Aviation logistics leaders must understand the concepts of battle command as discussed in FM 6-0. This requires that logistics Soldiers gain and sustain competency in executing individual and collective level combat tasks required for their unit and its associated OE.

5-31. Maneuver commanders must be willing to allocate combat power as an essential part of the mission to defend high risk aviation logistics units, and open and maintain ground and aerial LOCs as necessary. This may take the form of maneuver unit(s) escorting logistics convoy attaching a maneuver unit to reinforce perimeter defense, or occupying an area with sufficient force for a stated period of time to eliminate an air or ground threat.

5-32. The implied task for the aviation logistics commander is to possess the requisite skills to integrate the maneuver commander's forces into his security plan. All logistics leaders must be capable of defending an assigned AO by employing organic assets. As appropriate, the aviation logistics commander should coordinate with the aviation brigade for assistance in development of the area defense plan.

5-33. Aviation logistics leaders and their Soldiers must know how to execute the tactical enabling operations of road marches and tactical resupply convoy or LOGPACs.

LOGISTICS DOCTRINE FOR THE AVIATION BRIGADE

5-34. One goal of a transformed logistics system is to reduce reliance on stockpiles and static inventories located at each echelon that was characteristic of previous Army supply-based systems. In addition, the reduction of large stockpiles has assisted the accuracy of reporting by the user and logistician within their assigned STAMIS system.

5-35. This does not mean there are no on-hand supplies within the aviation brigade. For example, the unit has limited combat spares (comprised of authorized stockage list [ASL] items, prescribed load list (PLL), shop, and bench stock, as appropriate). Hence, once the request is submitted, it is expected to be satisfied in a timely manner. Use of the BCS-3 provides accurate and timely COP for logistics actions.

5-36. This type of logistics system combines a COP for logistics actions and its capabilities with efficient, yet effective delivery systems to form a seamless distribution pipeline. In essence, the supply pipeline becomes part of the warehouse representing inventory in motion, thereby reducing, but not eliminating, both organizational and materiel layering in forward areas.

5-37. Logisticians control the destination, speed, and volume of the distribution system. In-transit visibility (ITV), total asset visibility (TAV), advanced materiel management, and advanced decision support system technology provide logisticians with access and visibility over all items within the distribution pipeline. This visibility allows them to redirect, cross-level, and mass logistics assets more effectively in support of the commander's intent. Logisticians also maintain SU of the OE via the BCS-3, greatly facilitating planning and execution.

5-38. The logistics system relies on reduced order to receipt time to produce efficiency, but is designed with an overall intent to be effective in a combat environment. A goal of distribution-based logistics is direct throughput from the theater's sustainment brigade to the aviation brigade's ASB or, as needed, to the FSC or aviation maintenance company in the aviation battalion. Throughput distribution bypasses one or more echelons in the supply system to minimize handling and speed delivery to forward units. Improved materiel management systems allow supplies to be tailored, packaged into configured loads for specific supported units. This is based on a specific time and location point of need, and synchronized through distribution management channels based on the combat commander's mission and OPTEMPO.

5-39. Improved delivery platforms, such as the palletized load system and container roll in/roll out platform are used to deliver materiel to support units. Using ITV/TAV, delivery is tracked and managed from higher echelons to points as far forward as possible. Additional enablers include advanced satellite based tracking systems, movement tracking systems (MTSs), and radio frequency identification. Radar tracking station tags provide detailed distribution platform interrogation of items/material/stocks that, in turn, provide detailed asset visibility to the distribution system managers and forward units. This tracking much improves the materiel management system. BCS-3 greatly assists in this process.

5-40. Lastly, a secure intermediate staging base located in close proximity to the area of responsibility may be required to conduct rapid resupply when needed. All these aforementioned methodologies allow modular logistics units to focus on their supported units while conducting security operations.

ORGANIZATIONAL DESIGN OF THE SUSTAINMENT BRIGADE

5-41. A sustainment brigade conducts EAB replenishment operations to the aviation brigade. The sustainment brigade performs the functions of COSCOMs, DISCOMs, and ASGs.

5-42. The sustainment brigade has a command and staff structure capable of providing logistics management at the tactical and operational level. This includes providing an aviation brigade with external support to area support or the unassigned areas on a noncontiguous battlefield. In support of an aviation brigade, the sustainment brigade staff coordinates with the division Assistant Chief of Staff-Logistics (G-4) to plan and direct sustainment operations. Figure 5-1, page 5-8, illustrates the general structure of a sustainment brigade.

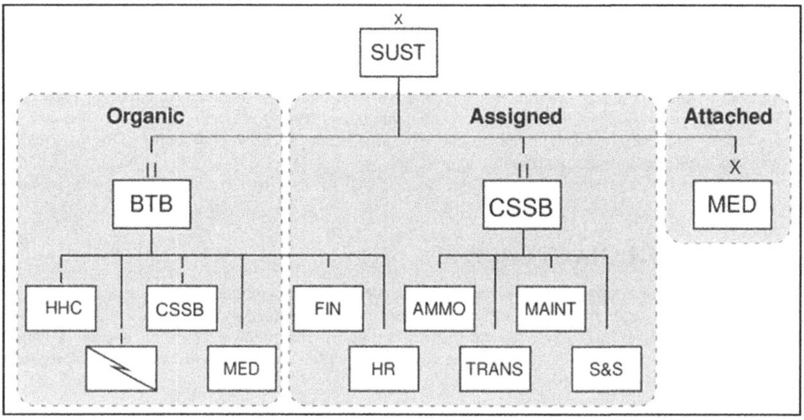

Figure 5-1. Example organization of the sustainment brigade

AVIATION BRIGADE LOGISTICS DESIGN

5-43. Preplanned pauses in battle rhythm allow combat forces to replenish routinely. Pulse operations include movement from the decisive operations zone to MSOs and redeployment to the decisive operations zone.

5-44. Aviation logistics organizations are designed to place the right logistics resources at the right location at the right time. Aviation logistics organizations primarily consist of an ASB within the aviation brigade at division and corps level, aviation support platoon at the theater level, and aviation maintenance company and FSC with each operational aviation battalion. These units collectively form the framework for aviation logistics in the Army's redesigned force structure.

5-45. In the area of logistics organization and C2, the following is true for the aviation brigade:

- The ASB commander and battle staff manage the change from legacy push to talk C2 systems to the digitized C2 architecture.
- The aviation brigade S-4, ASB, and battalion S-4s use the BCS-3 to achieve the LCOP.
- FSCs and aviation maintenance companies are organic to and logistics providers for aviation brigade battalions.
- The ASB is organized and equipped to support split-based sustainment and security operations.
- The ASB has a class IV storage capability at its ATP, which enables establishing an ammunition transfer holding point (ATHP).
- The ASB's ground maintenance capability primarily supports the ASB's assigned companies and aviation brigade headquarters. The ASB's ground maintenance capability provides limited or no back-up to the FSCs or aviation maintenance companies within the brigade's aviation battalions.
- The ASB has a Level I enhanced medical capability with its organic medical platoon.
- The ASB has a combat service support automation management office (CSSAMO) to support the aviation brigade logistics automation systems.

AVIATION BATTALION FORWARD SUPPORT COMPANY

5-46. An FSC is assigned to each operational aviation battalion and consists of a headquarters platoon, distribution platoon, and ground maintenance platoon. The FSC is designed to provide ground, air, missile, and above ground support equipment systems support; refueling and rearming support; and necessary logistics support. The FSC coordinates with the ASB for additional logistics as required. Each of the FARPs can be task organized to support continuous operations by providing support for maintenance, armament, and rearming and refueling. The FSC also maintains 2 days of supply (DOSs) of class I, provides field feeding and distribution support, maintains class IX (ground) repair parts, and conducts ground maintenance, while maintaining one combat load of class III (B) and class V for its supported battalion.

LOGISTICS CHARACTERISTICS

5-47. The Army has developed basic logistics fundamentals for supporting military operations. How well the combat commander emphasizes accurate and timely reporting and incorporates logistics leaders into the planning and preparing process impacts success or failure. The following paragraphs discuss logistics characteristics and methods of resupply. The commander on the ground must always plan and prepare for mission execution based on his own mission analysis.

5-48. The combat commander succeeds or fails by how well the logistics operators on the battlefield understand and adhere to the logistics characteristics as discussed in FM 4-0. The logistics characteristics are—

- Responsiveness.
- Simplicity.
- Flexibility.
- Attainability.
- Sustainability.
- Survivability.
- Economy.
- Integration.

PULSED LOGISTICS

5-49. Support that does not come in a continuous stream but arrives in distinct packages is called pulsed logistics. Pulsed LOGPACs include the support units, as well as engineers, air and missile defense, and combat units for security—a combined arms approach for logistics support. Pulsed logistics assist combat commanders in maintaining a high degree of combat power, while at the same time reducing the requirement on logistics units or their supported units to secure line of communications (LOC) at all times and in all places within the OE.

5-50. Pulse operations are used where division and corps operations allow for cycling of maneuver BCTs to temporary bases in which the brigade rests, refits, and receives large quantities of supplies. Hence, pulse operations are used so maneuver units pulse in and out of contact to be replenished and returned to the fight, or readied for another mission. Pulsed logistics are especially important when sustaining combat units widely distributed over a noncontiguous battlefield or a battlefield with LOCs that can only be secured temporarily.

METHODS OF DISTRIBUTION

5-51. Units use voice and digital means to request resupply and report status. The method used is determined after an analysis of the factors of METT-TC. The three distribution methods or resupply are—

- **Supply point distribution.** Supply point distribution requires unit representatives to move to a supply point to pick up supplies using their organic transportation.

- **Unit distribution.** Unit distribution provides delivery of supplies directly to the unit. A unit representative meets the resupply package at the logistics resupply point and guides the package to the unit's position. The ASB may use logistics convoys to conduct unit distribution operations.
- **Throughput distribution.** Shipments bypass one or more echelons in the supply chain and speed delivery forward. Throughput is more responsive to the user, provides more efficient use of transportation assets, and supplies are handled or transloaded less. Throughput to forward areas leverages configured loads, containerization, information, force structure design, technological enablers, and C2 relationships to deliver sustainment from the operational level directly to the customer or its supporting unit. Throughput is used frequently to resupply FARP operations.

Supply Operations

5-52. Supply operations involve acquisition, management, receipt, storage, and issue of all classes of supply except class VIII. FM 3-04.500, FM 4-0, joint publication (JP) 4-0, JP 4-03, and FM 10-1 give more details on supply operations.

Class I (Subsistence)

5-53. The class I supply system during the initial phase of an operation pushes rations. Personnel strength, unit location, type of operations, and feeding capabilities determine the quantities and types of rations pushed forward. As the battlefield stabilizes, the supply system converts to a pull system. Rations are throughput as far forward as possible.

5-54. The brigade S-4 generates ration replenishment requests for basic loads and monitors operational ration requests. Requests are based on personnel strength. Class I ration requests are consolidated by the battalion S-4 sections and forwarded to the aviation brigade S-4 or appropriate support area if operating independently. Extra rations usually are not available at distribution points; therefore, ration requests must accurately reflect personnel present for duty, including attached personnel. The brigade S-4 section draws rations from the distribution point and issues them to subordinate units.

Classes II, III (Packaged), IV

5-55. Classes II, III, and IV construction materials are handled in a manner similar to class I. Requisitions originate at the brigade unless a subordinate unit is operating under another headquarters. Normally, the materiel management center (MMC) authorizes shipment to the supply point in the support area via unit distribution. The items are then distributed to the battalions using supply point distribution. In some cases, the items may be throughput from the division, corps, or theater to subordinate battalions.

Class III (Bulk Petroleum, Oils, and Lubricants)

5-56. The basic load of class III bulk is the hauling capacity of the unit's fuel vehicles including the fuel tanks of unit vehicles. Topping off aircraft, vehicles, and equipment when possible, regardless of the fuel level, is essential to continuous operations.

5-57. Units normally use fuel forecasts to determine bulk petroleum, oils, and lubricants (POL) requirements. Battalions estimate the amount of fuel required based on projected operations, usually for the period covering 72 hours beyond the next day. Battalion S-4s forward requests through the brigade S-4 to the appropriate MMC. Units draw bulk POL from the support area class III supply point by unit distribution. Fuel trucks return to battalion areas either as a part of the LOGPACs or to refueling points in FARPs.

5-58. A key exception to this principle is refuel-on-the-move operations. Although these operations may use unit assets, typically they involve equipment of the supporting fuel unit. The purpose is to ensure the supported unit's vehicles and bulk fuel assets are topped before critical phases of an operation.

5-59. Class III bulk for the aviation brigade is delivered by sustainment brigade assets. The sustainment brigade can store a one-day supply of class III bulk. The fuel is stored and distributed from collapsible bladders or 5,000-gallon tanker trailers. Class III bulk normally is delivered to the ASB, and routinely delivered by the sustainment brigade as far forward as the aviation BSA. However, it may be delivered as far forward as FARPs in certain situations.

Class V (Conventional Ammunition)

5-60. Conventional ammunition is the standard ammunition associated with conventional weapons such as M240 machine-guns for the UH-60 and weapon systems mounted on the AH-64 and OH-58D. These classes include standard explosives such as hand grenades, claymores, C-4, and pyrotechnics (flares, star clusters, and smoke grenades). Special ammunition, which generally does not apply to the aviation brigade, includes nuclear ammunition, special missile warheads, and rocket motors.

5-61. Normally, the S-4 requests ammunition from the appropriate MMC. Ammunition managers use combat loads rather than DOSs. Combat loads measure the amount of class V a unit can carry into combat on its weapons system. Once the request has been authenticated, the ammunition is distributed to the battalion FSC by the ASB's distribution company.

Required Supply Rate

5-62. The RSR is the estimated amount of ammunition needed to sustain operations of a combat force without restrictions for a specific period. RSR is expressed in rounds per weapon per day. This RSR is used to state ammunition requirements. The S-3 in conjunction with the S-4 normally formulates the brigade RSR, but it is often adjusted by higher headquarters.

Controlled Supply Rate

5-63. The controlled supply rate (CSR) is the rate of ammunition consumption (expressed in rounds per day per unit, weapon system, or individual) that can be supported for a given period. It is based on ammunition availability, storage facilities, and transportation capabilities. A unit may not exceed its CSR for ammunition without authority from higher headquarters. The S-4 compares the CSR against the RSR; then remedies shortages by requesting more ammunition, sub-allocating ammunition, cross-leveling, or prioritizing support to subordinate units. The commander establishes CSRs for subordinate units.

Basic Load

5-64. Basic load is the quantity of ammunition authorized by the theater commander for wartime purposes and required to be carried into combat by a unit. The basic load provides the unit with enough ammunition to sustain itself in combat until the unit can be resupplied. The unit basic load (UBL) may not be the appropriate load to conduct operations based upon contingencies. Any deviation from the UBL is requested early for approval and resourcing.

Combat Load

5-65. Combat load is the quantity of supplies such as fuel or ammunition carried by the combat system or Solider into combat.

Class VI (Personal Demand Items)

5-66. Class VI supplies may be made available through local procurement, transfer from theater stocks, or requisitioning from the Army and Exchange Service. Available shipping space dictates class VI supply to theater. Class VI items are personal care items, candy, and other items for individual consumption. Health and comfort items (formally referred to as ration supplement sundry packages) are class VI supply items managed by the Defense Personnel Supply Center. They are issued through the standard supply system (normally class I supply channels) without cost to Soldiers in the early stages of a deployment. They contain items such as disposable razors, toothbrushes, toothpaste, and other personal care items. Defense Logistics Agency Regulation 4145.36 contains additional information on these packages.

Class VII (Major End Items)

5-67. Class VII supplies consist of major end items such as vehicles and aircraft. Because of their importance to combat readiness and high costs, class VII items usually are controlled through command channels and managed by the supporting MMC. Each echelon manages the requisition, distribution, maintenance, and disposal of these items to ensure visibility and operational readiness. Units report losses of major items through both supply and command channels. Replacement requires coordination among materiel managers, class VII supply units, transporters, maintenance elements, and personnel managers.

Class IX and Class IX (A) (Repair Parts)

5-68. The division MMC normally manages class IX. Within the battalions, the ASCs maintain bench stock and PLL. ASL items are maintained at the ASB level.

5-69. Class IX requisitions begin with the unit filling requisitions from its benchstock, shopstock, or PLL. If the item is not on hand or stocked or at zero balance, the requisition is passed to the supply support area (SSA). This unit fills the request from its ASL stocks or passes the requisition to the MMC. The ground maintenance sections of aviation units normally maintain class IX ASL for ground equipment.

SUPPORT BY HOST NATION

5-70. Logistics support and transportation may be provided by host nation organizations and facilities. Common classes of supply may be available and obtained from local civilian sources. Items may include barrier and construction materials, fuel for vehicles, and some food and medical supplies. Requisition and distribution are coordinated through logistics and liaison channels.

SECTION III – MAINTENANCE

PRINCIPLES

5-71. Maintenance is a combat multiplier. When enemy forces have relative parity in numbers and quality of equipment, the force that combines skillful use of equipment with an effective maintenance system has a decisive advantage. Such a force has an initial advantage in that it enters battle with equipment likely to remain operational longer. A subsequent advantage is it can repair damaged equipment, make it operational, and return it to the battle faster.

5-72. The maintenance system is organized around forward support. All damaged or malfunctioning equipment should be repaired onsite or as close to the site as possible.

SUPPORT SYSTEM STRUCTURE

5-73. The maintenance support system is a two-level structure—defined as field maintenance and sustainment maintenance. Field maintenance units concentrate on the rapid turnaround of equipment to the battle, while sustainment-level maintenance units repair and return equipment to the supply system.

Field Maintenance

5-74. Field maintenance is performed by aviation brigade personnel. Aviation battalions perform maintenance within their capability both in the flight companies and within their internal aviation maintenance companies. They are limited by sets, kits, outfits, and tools to keep them responsive and flexible, thus making them more agile. Battalions are authorized to perform unit maintenance detailed in the technical manuals (TMs) in accordance with AR 750-1. The ASC contained within the ASB is equipped with enhanced sets, kits, outfits, and tools to perform intermediate maintenance detailed in aircraft TMs. The ASC is authorized to perform intermediate maintenance in accordance with AR 750-1.

Sustainment Maintenance

5-75. Sustainment maintenance is performed within field repair activities, Army Depot, Aviation Classification and Repair Depot and original equipment manufactures either by contracted representatives or within their factories. On a case by case basis, the aviation brigade may obtain authorization via the assigned Aviation and Missile Command logistics assistance representative to affect repairs classified as depot in accordance with aircraft TMs. Army depots are often positioned at fixed bases within the CONUS. A graphic depiction of two-level maintenance, which also illustrates the relationship of field to sustainment maintenance, is shown in figure 5-2.

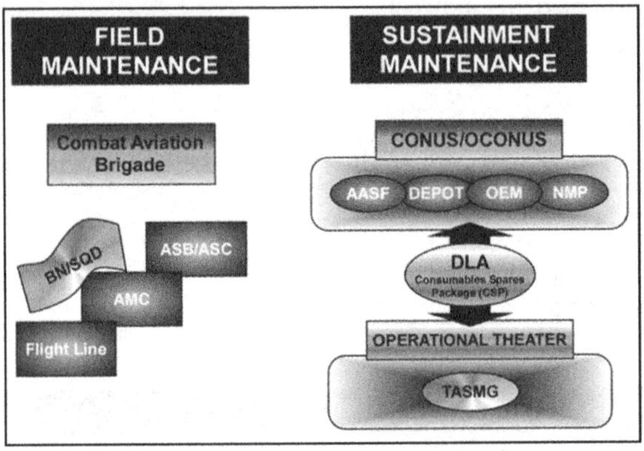

Figure 5-2. Two-level aviation maintenance and sustainment

VEHICLE AND GROUND EQUIPMENT MAINTENANCE AND RECOVERY OPERATIONS

MAINTENANCE SUPPORT STRUCTURE

5-76. Ground maintenance support for each battalion is provided by their organic FSC. The ASB's headquarters and support company (HSC) provides ground maintenance support to the aviation brigade HHC and ASB. Sustainment level units provide maintenance assistance as required.

PREVENTIVE MAINTENANCE CHECKS AND SERVICES

5-77. The operator or crew and organizational maintenance personnel perform unit maintenance that includes scheduled and unscheduled unit-level maintenance, repair, and preventive maintenance checks and services (PMCS). PMCS maintains the operational readiness of equipment through preventive maintenance and early diagnosis of problems.

FIELD MAINTENANCE

5-78. Field maintenance units are tailored to the weapons systems of the supported unit and provide maintenance with a multi-capable mechanic. They provide extensive maintenance expertise and component replacement and limited component repair. This level of maintenance is normally found in the HSC of the ASB and FSC of the aviation battalion.

SUSTAINMENT MAINTENANCE

5-79. Sustainment maintenance is characterized by extensive component repair capability. It repairs damaged systems for issue through the supply system as classes II, VII, or IX items. This level of maintenance is normally found at theater or depot level.

VEHICLE AND EQUIPMENT RECOVERY PROCEDURES

5-80. The recovery manager coordinates recovery operations with the overall repair effort to best support the commander's priorities and tactical situation. FM 4-30.31 describes the technical aspects of vehicle recovery operations.

Recovery Principles

5-81. When the unit recovers its equipment but lacks the physical means to recover an item, it requests assistance from the supporting maintenance element. Management of recovery operations is centralized at the battalion whenever possible.

5-82. Maintenance personnel repair equipment as far forward as possible within the limits of the tactical situation, amount of damage, and available resources. Recovery vehicles return equipment no further to the rear than necessary, usually to the maintenance collection point of the supporting maintenance unit.

5-83. Recovery missions that might interfere with combat operations, or compromise security, are coordinated with the tactical commander.

AVIATION MAINTENANCE OPERATIONS

5-84. The aviation maintenance company within each battalion provides unit maintenance above the capability of the flight companies. The ASC assigned to the ASB provides primarily intermediate maintenance and secondarily backup unit maintenance to the aviation brigade's battalions.

5-85. Aviation maintenance is performed on a 24-hour basis. Emphasis is on component replacement rather than repair. Such replacement requires increased stockage of line replaceable units (LRUs) and quick change assemblies. Damaged or inoperable aircraft requiring time-consuming repair actions are handled in more secure areas toward the rear. FM 3-04.500 provides more detail.

SCHEDULED MAINTENANCE

5-86. Commanders avoid situations that cause an excessive number of aircraft to require scheduled maintenance at the same time, or they avoid situations in which scheduled maintenance must be overflown. All imminent scheduled maintenance should be accomplished before deployment or initiation of surge operations. Refer to FM 3-04.500 for further information concerning scheduled maintenance flow.

PHASE AND PROGRESSIVE PHASE MAINTENANCE

5-87. Ongoing operations, training exercises, and deployments can have a major impact on readiness (for example, flying too many aircraft into scheduled maintenance at a critical time). OPTEMPO, deployments, training, and availability of resources (tools, maintenance personnel, repair parts, special equipment) must be considered when planning phase maintenance (AH-64, CH-47, and UH-60) and progressive phase maintenance (OH-58D) inspections. To facilitate phases in fast-moving operations, phases normally are done at the ASB.

UNSCHEDULED MAINTENANCE

5-88. Unscheduled maintenance or repair is generated by premature or unexpected malfunction, improper operation, or battlefield damage. Units must be doctrinally and organizationally prepared to apply responsive corrective action on an as-needed basis. Maintenance support teams (MSTs) must be identified prior to missions and assigned to scheduled shifts to quickly react to unscheduled maintenance requirements, ensuring aircraft availability for follow-on missions.

BATTLEFIELD MANAGEMENT OF DAMAGED AIRCRAFT

5-89. BDAR/recovery operations are planned and coordinated in detail in conjunction with PR operations to minimize risk. Recovery operations are those that move an aircraft system or component from the battlefield to a maintenance facility. Recovery may require on-site repair for a one-time flight, or movement by another aircraft or surface vehicle. In extreme circumstances, only portions of inoperative aircraft may be recovered. An aircraft will be cannibalized at a field site only when the combat situation and aircraft condition are such that the aircraft would otherwise be lost to enemy forces. Refer to FM 3-04.500 and FM 3-04.513 for more detailed information on aircraft recovery.

AVIATION LIFE SUPPORT SYSTEM

5-90. Commanders ensure mission-required aviation life support equipment (ALSE) is on hand in sufficient quantities, and is in serviceable condition. Commanders are required to establish an aviation life support system maintenance management and training program budget to meet resource requirements. Funding for equipment, supplies, and repair parts is imperative. When preparing the budget, review AR 95-1, common table of allowances (CTA) 8-100, CTA 50-900, and applicable MTOEs and tables of distribution and allowances.

SAFETY DURING MAINTENANCE OPERATIONS

5-91. An effective safety program for maintenance operations is a basic requirement in all units. Everyone must be alert to immediately recognize and correct potentially dangerous situations. Accidents can cause more losses than enemy action unless safety is embraced by the unit.

ACCIDENT CAUSES

5-92. An accident in the shop, FARP, or air is seldom caused by a single factor such as human error or materiel failure. Accidents are more likely to result from a series of contributing incidents. The following areas require constant command attention to prevent aviation accidents:

- Human factors.
- Training, education, and promotion.
- Equipment design, adequacy, and supply.
- Normal and emergency procedures.
- Maintenance operations.
- Work Environment.

5-93. More complex aircraft have higher maintenance-related mishap rates. Commanders and maintenance supervisors must ensure their personnel learn from maintenance errors generated in their own units. Flightfax and other publications provide additional examples and information. All personnel must strictly adhere to published maintenance procedures and apply CRM at all levels of operations.

SAFETY REGULATIONS

5-94. AR 385-10 regulates overall safety. AR 385-10 regulates the Army aviation accident prevention program. Department of the Army pamphlet (DA Pam) 385-40 and AR 385-10 cover accident investigation and reporting.

RESPONSIBILITIES

5-95. The quality assurance (QA) section has primary responsibility of safety for all maintenance work performed on aircraft or their components. However, everyone in the unit has responsibilities in the unit's maintenance safety and aviation accident prevention programs. General responsibilities for key personnel are outlined in the following paragraphs. Appendix D contains additional information.

Unit Commander

5-96. Commanders ensure all unit activities are conducted according to established safety rules and regulations. These regulations include AR 385-10, DA Pam 385-40, FM 5-19 and local directives. Commanders also determine the cause of accidents and ensure corrections are made to prevent recurrence. When deviation from an established safety rule is desired, commanders obtain permission from the appropriate higher commander.

Leaders

5-97. Effective supervision is key to accident prevention. Supervisors must apply all established accident prevention measures in daily operations. They should frequently brief subordinates on safety procedures, get suggestions for improving safety practices, and announce any new safety procedures. Recommended agenda items are listed below.

- Overall job and expected results.
- The how, why, and when of the job, and any ideas from the group on ways to improve methods and procedures.
- Part each person contributes.
- Existing and anticipated hazards and action needed to resolve these problems.
- The need for prompt, accurate reporting of all injuries, accidents, or near accidents.
- Basic first aid procedures, training, and readiness.
- The need to search constantly for, detect, and correct unsafe practices and conditions to prevent accidents and injuries.

Individuals

5-98. All personnel must be aware of the safety rules established for their individual and collective protection. Each person must read and follow unit SOPs, instructions, checklists, and other safety-related information. They must report safety voids, hazards, and unsafe or incomplete procedures. Each Soldier must follow through until the problem is corrected.

SECTION IV – AVIATION SUSTAINMENT UNITS

FLIGHT COMPANY

5-99. Crew chiefs perform aircraft launch and recovery operations, and maintain aircraft logbooks in accordance within Army guidance and unit SOPs. They perform both scheduled and unscheduled unit maintenance to include replacement of major subsystem components, maintenance operational checks, and main and tail rotor vibration analysis. The battalion flight companies receive backup support from the aviation maintenance company to perform both scheduled and unscheduled maintenance. Refer to FM 3-04.500 for additional information.

AVIATION MAINTENANCE COMPANY

5-100. The aviation maintenance company is comprised of three modular aviation support platoons including the headquarters platoon, airframe repair platoon, and component repair platoon (CRP). The purpose of the aviation maintenance company is to repair and maintain aircraft. The company is organized to provide quick, responsive, real-time internal maintenance support and repair within its capability.

5-101. The aviation maintenance company troubleshoots airframe and component malfunctions and performs maintenance and repair actions requiring less than 2 days to complete. The aviation maintenance company is authorized to perform maintenance at the unit level in accordance with the maintenance allocation chart (MAC). It conducts BDAR and DART within its capability. During operations, most aviation platoons or companies are in the forward portion of the support area.

5-102. The aviation maintenance company provides mobile, responsive support through MSTs used to repair aircraft onsite or prepare them for evacuation. The aviation maintenance company commander and production control officer coordinate and schedule maintenance at forward locations of the battalion. Members of the forward element must be able to diagnose aircraft damage or serviceability rapidly and accurately. MSTs follow these principles:

- Teams may be used for aircraft, component, avionics, or armament repair.
- When time and situation allow, teams repair on site rather than evacuate aircraft.
- Teams must be 100 percent mobile and transported by the fastest means available (normally by helicopter).
- Teams sent forward must be oriented and equipped for special tasks.

5-103. In some situations, normal maintenance procedures must be expedited to meet operational objectives. In such cases, the unit commander may authorize use of aircraft combat maintenance and BDAR procedures. Aircraft combat maintenance and BDAR is an aviation maintenance company responsibility with backup from supporting ASB units. The concept uses specialized assessment criteria, repair kits, and trained personnel to return damaged aircraft to the battle as soon as possible. Often, these repairs are only temporary. Permanent repairs may be required when the tactical situation permits. This method is used to meet operational needs. It is not used when the situation allows application of standard methods.

HEADQUARTERS PLATOON

5-104. The headquarters platoon is comprised of four sections—headquarters; production control; QA; and technical supply. This platoon provides internal management, quality of repairs, and logistics support within the battalion. The technical supply section operates logistics STAMIS, requisitions class IX (A) spares and manages the battalion PLL. Oversight is provided by the battalion aviation material officer assigned to the S-4.

AIRFRAME REPAIR PLATOON

5-105. Airframe repair platoons (ARPs) assigned to an aviation maintenance company provide their supported aviation units with scheduled and unscheduled maintenance support. Primary responsibility for unscheduled maintenance falls on the owning unit. However, when unit OPTEMPO increases, unscheduled maintenance support can be coordinated and requested through the aviation maintenance company production control office. If the line company cannot complete the unscheduled maintenance in 1 day or less, it should contact the production control office and request airframe repair platoon (ARP) maintenance support. Location of the maintenance action can then be further coordinated by the line company and the production control section. Primary responsibility for periodic scheduled maintenance falls upon the owning unit. Prolonged scheduled maintenance—including aircraft phases, compliance with recently published aviation safety action messages/technical bulletins—can lead a supported unit to request maintenance support from the aviation maintenance company. Maintenance support can be coordinated and requested by the owning unit through the production control office.

COMPONENT REPAIR PLATOON

5-106. The CRP is assigned a headquarters section, shops section, and a systems repair section. The CRP contains assigned aviation repair specialty military occupational specialties (MOSs) to include avionics, armament, powerplant/powertrain, hydraulics, pneumatics, and sheet metal repair assets. The CRP diagnoses airframe and component malfunctions and performs maintenance, repair actions, and removes and installs LRUs within its capabilities.

5-107. The shops section contains an armament/avionics/electrical repair team. The armament team is responsible for troubleshooting and repairing armament systems, subsystems, and components. Personnel assigned to the armament systems repair team conduct preventive maintenance and conduct testing and troubleshooting of aircraft weapons systems and subsystems. These personnel also perform cleaning, servicing, and ammunition loading and unloading of weapons systems to include configuration changes. The armament team is responsible for repairing and replacing weapons platforms components in accordance with applicable publications.

5-108. The CRP systems repair section performs preventive maintenance of aircraft components and structures that require specialized technical skills. In addition, maintainers assigned to this section perform scheduled and unscheduled maintenance, troubleshoot faulty components, remove and replace aircraft components, perform BDAR procedures and manage assigned sets, kits, and outfits at the platoon level, and provide mission support to flight companies.

AVIATION SUPPORT BATTALION

5-109. The ASB (figure 5-3) is the primary aviation logistics organization above aviation battalion. The ASB is organic to the aviation brigade and provides all logistics functions necessary to sustain the aviation brigade during full spectrum operations. It consists of four companies—the HSC, distribution company, NSC, and ASC. The ASB provides aviation and ground field maintenance, network communications, resupply, and medical support. The HSC provides medical support and conducts ground field maintenance and recovery. The distribution company functions as a SSA and distributes supplies to subordinate units of the aviation brigade. The NSC provides network and signal support to the aviation brigade headquarters. The ASC provides field level maintenance, to include intermediate level maintenance, and support for on-aircraft and critical off-aircraft maintenance of UAS. The ASC also conducts BDAR and provides backup support to the aviation maintenance companies.

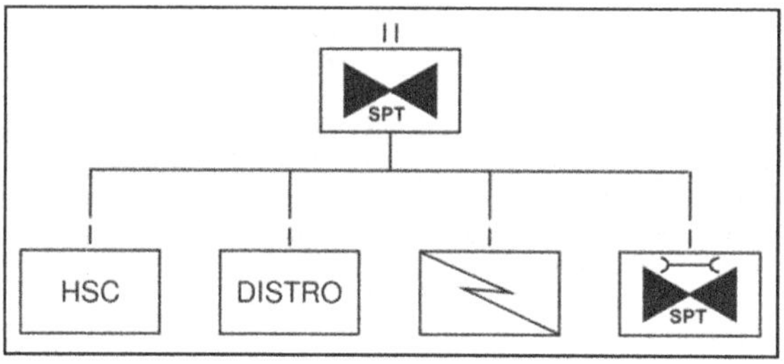

Figure 5-3. Aviation support battalion

MISSION

5-110. The ASB distributes supply classes I, II, III, IV, V, VIII, and IX. It performs field maintenance and recovery (air and ground), and possesses the HSS assets to conduct force health protection Level I enhanced for the aviation brigade. The ASB carries logistics stocks that exceed the organic carrying capability of the aviation brigade battalions that are generally one DOS for most classes of supply except classes III (B) and IV where it is one combat load for the brigade. FSCs have the same type of carrying capacity relative to the support of their battalion. The ASB plans and coordinates for the aviation brigade's logistics requirements in coordination with the brigade S-4 during the brigade's MDMP. The ASB executes replenishment operations for the FSCs and aviation maintenance companies in concert with the

OPLAN developed by the brigade. The ASB is the parent battalion headquarters for the NSC in support of brigade headquarters.

TASKS

5-111. ASB tasks are—

- Conduct field maintenance, both ground and air, to include UAS and recovery operations.
- Provide signal and network support for the aviation brigade headquarters to enable C2 of subordinate battalions and the aviation brigade or ABTF.
- Provide logistics for ground, air, missile, and above ground support equipment systems.
- Monitor and update the current situation such as developing logistics and tactical COPs.
- Plan, synchronize, manage, and execute sustainment operations at brigade level within the aviation commander's battle rhythm.
- Plan, establish, maintain, and synchronize distribution management operations within the aviation brigade; link back with the division for coordination of requirements and synchronization of flow.
- Determine and anticipate logistics requirements for maneuver operations.
- Provide Level I enhanced medical support plus emergency resuscitative surgery. Plan, coordinate, and provide emergency medical treatment and advanced trauma management for wounded and disease and nonbattle injury patients and sick call services.
- Provide mass casualty management to include triage, treatment, and evacuation.
- Integrate mission tailored logistics augmentation to support the concept of maneuver as required.

LIMITATIONS

5-112. The ASB is not designed to provide the following logistics functions:

- Medical support is limited to an enhanced Level I medical platoon.
- Field services.
- Mortuary affairs (planning only). No collection, processing and evacuation without augmentation.
- Laundry and bath is not organic at this level. Support is provided by the sustainment brigade.
- Limited financial management.
- Limited class VIII/IX storage capability.
- Limited capability to reconfigure load. Ammunition from EAB must be in strategic or operational configured loads.
- No fire fighting capability.
- Explosive ordnance disposal is provided by the maneuver enhancement brigade.
- Human resources other than its own unit S-1 human resource operations. Relies on the sustainment brigade to provide additional critical wartime personnel support.
- Legal support is limited to the assigned BOLT; augmentation to support all Judge Advocate General functions are required.
- There is no built-in ground maintenance back-up support to the maneuver units; however, the ASB commander can provide support when capacity is available.
- No optical fabrication and blood product management support.
- No organic aeromedical evacuation support. Aeromedical evacuation support is organic to the aviation brigade.

HEADQUARTERS AND SUPPORT COMPANY

5-113. The HSC contains a typical logistics battalion staff structure with a command section, S-1, consolidated S-2/S-3, S-4, UMT, S-6, and support operations section. The battalion headquarters provides C2 and administration support for all organic and attached ASB units. The battalion headquarters also plans, directs, and supervises logistics support for the aviation brigade. The ASB has an organic CSSAMO that provides support for the entire brigade's logistics network and system, including unit level logistics systems–aviation (ULLS-A).

5-114. The support operations section is organized to coordinate logistics support and provide distribution management to the aviation brigade. The support operations section is also manned to accomplish contracting, medical and medical logistics, petroleum, ammunition, movement control, transportation, and mortuary affairs functions.

Support Company

5-115. The support company provides ground maintenance, medical, supply, and food service support to units organic and attached to the ASB.

5-116. The maintenance platoon is responsible for field level maintenance for all of the ASB's organic ground equipment.

5-117. The medical platoon provides Level I enhanced medical care. The platoon includes a flight surgeon, physician's assistant, health care NCOs, and health care specialists. The platoon is organized into headquarters, treatment, and evacuation sections. Additionally, the platoon has four ambulances. The brigade HHC and flight battalions retain their organic flight surgeons and medics in their organic medical treatment team. The medical platoon provides the following capabilities for the ASB:

- Emergency medical treatment and acute trauma management for wounded and disease and nonbattle injury patients.
- Sick call services.
- Ground ambulance evacuation from supported units.
- Mass casualty triage and management.
- Limited patient decontamination.

DISTRIBUTION COMPANY

5-118. The distribution company provides the aviation brigade a single source for all supply (less class VIII) and transportation operations. The distribution company includes a fuel and water platoon, supply platoon, and transportation platoon.

5-119. The fuel and water platoon has the capability to store and distribute 105,000 gallons (one DOS) of fuel for the brigade using three load-handling system modular fuel farms. Additionally, the platoon has the capability to set up and run multiple refuel points for brigade aircraft. The fuel and water platoon also has the capability to purify 30,000 gallons of water daily and can store 18,000 gallons of water. The platoon has an organic quartermaster petroleum QA team assigned to provide QA testing for bulk aviation fuel. The team performs quality evaluation and provides technical assistance for handling, storing, sampling, and identifying petroleum products and their containers for the aviation brigade.

5-120. The supply platoon has an SSA and ammunition transfer and holding point section. This platoon provides classes II, III (P), IV, V, VI, VII and IX DS to the brigade. The supply platoon receives, stores (limited), and issues classes II, III (P), IV, and IX. It also receives and distributes classes I and VI under the distribution based doctrine of pushing supplies to the FSCs and aviation maintenance companies, and receives and issues class VII as required. The platoon also maintains the classes II, III (P), IV and IX ASL for the brigade. The ATHP section supports the brigade with class IV and operates the brigade ATHP.

5-121. The transportation platoon's purpose is to add organic transportation and distribution capability to the brigade and increase mobility of the ASB. The transportation platoon also has the ability to transport classes V and IX to the supported FSCs and ASCs.

Network Support Company

5-122. The NSC provides 24-hour operations supporting the aviation brigade network. It provides signal elements designed to engineer, install, operate, maintain, and defend the network. It extends defense information systems network services to the brigade and its subordinate elements and provides basic network management capabilities. The company executes its technical mission under functional control of the brigade S-6 based on brigade OPORDs or other directives. The S-6 directs actions and movement of signal elements in support of brigade operations. The NSC commander maintains command authority over the company's assigned operational platoons or attached elements. Refer to appendix B for additional information.

Aviation Support Company

5-123. Aircraft maintenance above aviation battalion level is provided by the ASC of the ASB. The ASC is comprised of three platoons—headquarters, ARP, and CRP. Modularity within the ASC is based on a contact support team concept and uses five shop equipment contact maintenance vehicles per platoon. The ASC is capable of supporting brigade split based operations and appropriate support to ABFTs. The ASC primarily performs intermediate maintenance in accordance with the MAC; however, it also provides backup unit maintenance in support of aviation battalions. The ASC provides aviation logistics support operations for aviation brigade assets. It provides aviation and ground equipment maintenance, in a sustained combat environment, to include UAS and air traffic control equipment. The ASC also performs production control and quality control, conducts maintenance management, and provides maintenance test pilot (MP) functions. Additionally, ASCs have six-man electro-optics test facilities augmentation teams assigned.

Headquarters Platoon

5-124. The headquarters platoon contains the production control and QA sections and technical supply section. This platoon provides internal management of repairs, and quality of repairs and logistics support within the battalion. The tech supply section operates logistics STAMIS, requisitions class IX (A) spares and manages the battalion PLL. Oversight is provided by the battalion aviation material officer assigned to the S-4.

Aircraft Repair Platoon

5-125. The ARP performs maintenance actions which require more than 3 days to complete (such as phase maintenance and preventative maintenance and services). The ARP performs in depth troubleshooting and diagnosis of airframe and component malfunctions; fixes and fuels organic battalion equipment, ground vehicles and aviation general support equipment (GSE); operates and performs field level maintenance on aviation ground power units, generators and GSE; and performs BDAR. The ARP contains modular maintenance contact teams to support battalion level deployments (five sections/one per battalion). The primary methods of returning aviation systems to a mission capable status for a field level maintenance activity are through use of repair parts, BDAR, controlled substitution, controlled exchange, and class VII replacement. As the senior logistician in the brigade, the ASB commander tailors the ARP to support multiple ASCs and the aviation brigade's mission. Each ARP is assigned to a supported aviation battalion. In addition, component repair organizations may be attached to the ARP to facilitate rapid turnaround of critical sustainment level off aircraft tasks/components. The ARP is designed to provide on aircraft and critical off aircraft aviation field level maintenance, both unit and intermediate, in accordance with the MAC. The ARP also performs BDAR for all assigned aircraft and UAS in the aviation brigade. Long-duration low-frequency services such as phases are accomplished at the ARP.

Component Repair Platoon

5-126. The CRP repairs LRU components to TM standard and returns them to the user. The CRP fixes the night vision goggle (NVG) systems, ALSE, and avionics-electrical and hydraulic components to TM standard. It has limited capability to fabricate hydraulic lines and perform engine repair, prop and rotor repair, and armament and armament sub-system repair. The CRP also provides limited fabrication capability using welding and machine shops and operates intermediate level STAMIS. It also repairs and troubleshoots unit level STAMIS with support from CSSAMO in the HSC.

COMMAND AND CONTROL

5-127. ASBs use echeloned C2 to plan and direct operations. Battle command is tailored to meet the requirements of each operation. The battalion command group normally operates within the aviation brigade's OE as appropriate to meet the sustainment and force protection requirements. It consists of the commander and those selected to assist in controlling the operational and sustaining elements of the battalion. The commander determines the composition, nature, and tasks of the command group based on METT-TC analysis. As a minimum, the command group—

- Integrates support battalion and attached logistics assets in support of sustainment operations and signal operations.
- Controls sustainment operations and force protection operations.
- Maintains SU.
- When not at the main CP, the command group provides close situation information to the main CP of observations achieved by proximity to the activity.

5-128. The commander, SPO, and S-2/S-3 monitor the battle, develop the situation, analyze COAs, and control the companies except as noted for the NSC. The ASB commander's C2 structure for replenishment and security operations, and the systems that assist the commander, are used to see the battlefield and lead the battalion as it conducts operations.

5-129. The commander and S-2/S-3 are the only battalion level leaders that can issue tasking orders. The SPO should coordinate requirements and provide WARNOs for replenishment operations in support of the brigade, but the S-3 issues the tasking order.

5-130. The ASB commander's C2 structure for logistics centers on three entities—the ASB commander's location, his command group, and ASB CP. The logistician's headquarters enables the commander to maximize command, control, and IM for logistics and TACOPS. The ASB CP employs the current battle command systems required to C2 the ASB's organic and supporting Army WFFs and units. The ASB commander's C2 systems also enhance the logistics staff's ability to provide the commander with timely information; maintain an accurate COP; efficiently process, analyze, and disseminate battlefield information; and provide updated mission orders rapidly. The commander operates independently, establishing BCOTM as necessary, or operates from the ASB CP based on the situation and phase of the operation.

5-131. ASB C2 consists of key personnel, equipment, and CP from which the battalion commander, assisted by the battle staff, directs operations and sustains the force. METT-TC dictates the organization of C2 personnel, facilities, and location of the command group.

5-132. The duties of the ASB commander and support operations section are different enough from aviation battalion functions that an explanation of their responsibilities is warranted. The other functions of the battalion staff are generally similar to the Army's operational aviation battalions.

Commander

5-133. The ASB commander is the senior logistician for the brigade. He or she manages logistics through use of an array of digital information systems and a technologically competent battle staff capable of capitalizing on all other technological innovations. The ASB commander directs all units organic or attached to the battalion in support of the brigade's mission. He or she also has control of all elements in the aviation BSA for security and terrain management. He or she provides subordinate elements with clear missions, taskings, and a statement of his intent.

5-134. The battalion commander provides aviation maintenance and distribution management at the brigade level and maintains SU of the logistics assets required to support the brigade's responsibilities. The ASB commander's responsibilities include leadership, discipline, tactical employment, training, administration, personnel management, supply, maintenance, communications, and logistics activities of the battalion. The ASB commander's duties include—

- Establishing his CCIR and EEFI.
- Understanding capabilities and limitations of the battalion's personnel and equipment in performing the logistics mission to include security operations as well as those of logistics elements attached to him or her.
- Developing and providing a LCOP in meaningful terms for the brigade commander and his staff.
- Staying personally involved in and apprised of replenishment operations and the tactical situation throughout the brigade AO and BSA OE.
- Being proficient in the tactical employment of the battalion and its assigned and attached logistics elements.
- Establishing an effective perimeter defense plan for all assets within the BSA fully coordinated with the brigade S-3. Personally ensure establishment of the plan by subordinate commanders/leaders with on-site inspections.
- Developing fully coordinated, effective combat convoy movement plans with the brigade commander and his staff for execution if necessary with combined arms forces.
- Understanding the full capabilities of the tactical and logistics radio and data transmission capabilities available to the commander and his staff.
- Maintaining contact with higher, lower, and adjacent supported and supporting units. A liaison should be used if that is the best solution.
- Ensuring connectivity of STAMIS and FBCB2 with the brigade and supporting units.
- Knowing the responsibilities and capabilities of higher, lower, and supporting units and knowing the support required and what support each level or type of organization can provide.
- Using effective oral communications and writing clear directives and orders. For example—
  - Providing commander's intent and mission guidance.
  - Reviewing battle staff estimates of the tactical and logistics situation, their COA analysis, and then recommending the COA that best supports the brigade mission by sustaining the fighting capability of the brigade.
  - Stating his estimate of the situation and announcing his decision.
- Being familiar with the law of land warfare with respect to civilians, civil affairs, and CMO.
- Ensuring there is a well-known and rehearsed plan of command succession.

Commander's Location on Battlefield

5-135. Commanders consider their position in relation to the units they command and the mission. Their location can have important consequences for executing sustainment operations. Modern information systems can help commanders command throughout the AO without losing access to information and analysis of the CPs. Should commanders require a larger facility to temporarily exercise C2, a subordinate tactical CP can be used to establish communications to their CP.

5-136. At battalion level logistics, the ASB CP is normally the focus of information flow and planning. Yet the logistics commander cannot always visualize the battlefield and direct and synchronize operations from there. He or she must sometimes assess the situation on the ground—face-to-face with subordinate commanders and their Soldiers. Commanders design their C2 systems to position them where they can best command without losing the SU that allows them to anticipate situations and respond to opportunities and changing circumstances.

5-137. At the ASB level, commanders lead more indirectly through their subordinates. Commanders may want to have personal contact with or intervene to make decisions at the location or with the command executing the decisive operation. Similarly, when commanders lose sight of the situation, they need to reestablish a COP to achieve clear SU.

Support Operations Section

5-138. This section, under direction of the SPO, provides centralized, integrated, and automated C2 and planning for all distribution management operations within the battalion. It coordinates with logistics leaders, staff planners, and medical personnel in the fields of supply, maintenance, force health protection, mortuary affairs, and movement management for the support of all units assigned or attached in the brigade area. Its primary concern is supported units and increasing the responsiveness of support provided by subordinate units. It continually monitors support and advises the battalion commander on the ability to support future TACOPS. With the GCCS-A, BCS-3, FBCB2, and MTS, the support operations section has access to and receives information in near real time. Therefore, the support operations section possesses the capability to view the LCOP and combat power in the maneuver units allowing quick identification of problems to allocate resources more efficiently. The BCS-3 gives support operations the visibility of the logistics status from the ASB back to the sustainment brigade and potentially throughout the world depending on the level of detail required.

5-139. The support operations section serves as the point of contact (POC) for supported units. It directs problems to appropriate technical experts within subordinate branches. The duties of the support operations section include the following:

- Conducts continuous brigade focused LBP.
- Plans and coordinates for aerial resupply and plans for LZs in the vicinity of the BSA.
- Develops the logistics synchronization matrix.
- Submits logistics forecasts to the division sustainment brigade.
- Manages all flatracks throughput to and retrograding from the BSA.
- Coordinates and provides technical supervision for the ASB's sustainment mission, which includes supply activities, maintenance support, force health protection, and coordination of transportation assets.
- Identifies tentative force structure and size to be supported.
- Coordinates preparation of the support operations estimate on external support.
- Provides support posture and planning recommendations to the ASB commander.
- Sets up and supervises the logistics operations center located in the ASB CP.
- Coordinates with brigade S-3 air routes for supply and aeromedical evacuation support.
- Provides centralized coordination for units providing support to the brigade.
- Analyzes the impact of BCS-3 reports.
- Advises the battalion commander on the status of logistics support.
- Coordinates logistics support for units passing through the brigade's area. Works with ASB S-3, aviation brigade S-3, as appropriate, for terrain management and movement across other unit's AO.
- Analyzes contingency mission support requirements.
- Revises customer lists (as required by changing requirements, workloads, and priorities) for support of TACOPS.
- Coordinates external logistics provided by subordinate units.
- Advises the battalion commander on supportability of ASB support missions and of shortfalls impacting mission accomplishment.
- Serves as the single point of coordination for supported units to resolve logistics support problems.
- Plans and coordinates contingency support.
- Develops supply, service, maintenance, and transportation policies including logistics synchronization and maintenance meetings.

5-140. The SPO performs functions as the BCS-3 manager. The SPO must work in conjunction with the S-2/S-3, S-4, and S-6 to establish and manage the BCS-3 network and database. The SPO must maintain supply point and maintenance data entered into the system.

Supply and Services Cell

5-141. The support operations supply and service officer plans and recommends the allocation of resources in coordination with the supported chain of command, including coordination with the distribution section. The supply and service officer also forecasts and monitors distribution of supplies within the brigade. Information is entered into BCS-3 at the brigade S-4 and transferred to BCS-3 at support operations. This allows support operations to identify problems quickly and allocate resources more efficiently. The supply and service officer is responsible for mortuary affairs activities carried out within the brigade AO. He or she is also responsible to coordinate and monitor all transportation movements of replenishment stocks and services for and within the ASB.

5-142. The supply and services cell has two traffic management coordinators assigned to control movement of transportation assets in and around the ASB. The traffic management coordinators—

- Monitor, control, and supervise movement of personnel, equipment, and cargo.
- Develop and review movement programs (to include convoy planning) for logistics support functions within the ASB/BSA.
- Advise in preparation of support plans where transportation is required.
- Verify accuracy of movement control documents.
- Ensure allocation of transport capability is appropriate to accomplish each mission in a cost-effective manner.
- Coordinate support with the movement control office in the sustainment brigade's support operation section when transportation requirements exceed the ASB's capability.
- Anticipate and recommend use of main supply routes (MSRs) to the movement control office.

5-143. The addition of new enabling technologies allows the traffic management coordinators to track, trace, and divert transportation platforms operating in the brigade AO. The traffic management coordinators are responsible for the ITV in the theater of operations. ITV is best accomplished by the ASB movements NCO interfacing with other STAMIS to develop inbound/outbound requirements. ITV uses the MTS and other ITV technology to get a near real-time location of transportation assets and supplies. In addition, the traffic management coordinators are able to synchronize the delivery schedule via FBCB2 with customer units to minimize the offload/upload times. With FBCB2 and the MTS control station, the traffic management coordinators are now able to give specific coordinating instructions to vehicle operators without having to rely on manned control points. These new technologies allow information to be transferred between the brigade S-4, battalion S-4, ASB support operations section and the traffic management coordinators to schedule and synchronize transportation requirements within or in support of brigade/battalion operations.

Maintenance Cell

5-144. The support operations maintenance officer (MO) plans and recommends the allocation of resources in coordination with the supported unit's chain of command, including coordination of maintenance company operations. The support operations MO also forecasts and monitors the workload for all equipment by type. The MO and maintenance NCO use standard Army maintenance system-level 2 (SAMS-2) to collect and process maintenance operations data and assist in the management of maintenance operations. SAMS-2 processes maintenance information required to control workload, manpower, and supplies. The SAMS-2 capabilities are designed to assist in both maintenance and readiness management.

5-145. The aviation battalions transmit logistics SITREPs electronically to the brigade S-4 and ASB SPO. This allows support operations to identify problems quickly and allocate resources more efficiently. FBCB2 also provides map graphics that portray unit locations, grid coordinates, and terrain features so support operations can track maintenance on the battlefield.

5-146. The support operations maintenance cell develops plans and policies for repairable exchange and class IX operations. It monitors shop production and job STATREPs in the field maintenance company and FSCs. It also monitors and reviews combat spares and coordinates critical parts status with the sustainment brigade. For unserviceable items, the standard Army retail supply system box in the distribution company generates disposition instructions based on commander's guidance. Instructions include evacuation, controlled exchange, and controlled exchange policies. With the brigade S-4, the support operations maintenance cell reviews backlogs on critical weapon systems. For any additional support requirements, the ASB support operations section coordinates through the division materiel management branch.

Health Service Support Cell

5-147. For brigade force health protection operations, the HSS cell provides input to the brigade surgeon section (BSS) for inclusion into the force health protection annex of the brigade OPLAN. Refer to FM 4-02.21 for additional information on the BSS. The health service support officer provides BSS information on all medical activities to include attachment of sustainment brigade medical elements, class VIII resupply, MEDEVAC, and priority of force health protection for the BSA and brigade AO. The HSS cell plans for the use of nonmedical platforms for CASEVAC, and the support operations section manages their use during mass casualty operations.

5-148. The medical communications for combat casualty care system assist the force health protection cell and BSS in performing their responsibilities through the collection, integration, and transmission of medical information. These sections have near real-time information on the status of medical units, brigade unit medical readiness information, CASEVAC, medical supplies, and medical treatment.

SECURITY OF AVIATION SUPPORT BATTALION ASSETS

5-149. Sustainment elements locate most of their assets in the aviation BSA. These elements deploy in a manner that maintains unit cohesion, supports integration into the defensive plan for collocated units, and provides responsive support to the battalions. FM 3-04.500 contains additional information about logistics movement and operations.

5-150. The brigade commander's goal is to retain overall freedom of action for fighting military operations. This means MSRs are clear, unobstructed, and secure; units can move quickly and in an orderly fashion throughout the brigade area; logistics resupply via logistics convoys and reconstitution are sustained; and all sustainment and logistics units are secure.

5-151. The brigade commander is responsible for plans and operations throughout the brigade AO. He or she assigns tasks to subordinate and supporting commanders to accomplish all brigade missions. The brigade S-3 includes detailed planning for the ASB AO as part of operational planning for offensive, defensive, and sustainment operations and support operations missions.

5-152. The ASB commander is responsible for defense of the BSA. Hence, the BSA's perimeter defense is under C2 of the ASB commander. The ASB commander's plan of action must achieve adequate protection to ensure accomplishment of missions by BSA elements with as small a force as necessary, since any drain of time and personnel from operational activities adversely affect accomplishment of their missions. However, survival of the ASB is most important for the continuing success of the aviation brigade's ability to sustain itself.

5-153. The security planning starts with the aviation brigade's first WARNO during the brigade MDMP; whereas the sustainment concept of the support plan the ASB executes is developed during the brigade's MDMP. The ASB, like all other subordinate battalions of the brigade, conducts security operations planning for itself and the units within the BSA.

5-154. The ASB commander has control of all elements in the BSA for defense and positioning. The major elements in the BSA assist with forming the perimeter in a contiguous manner. It is possible due to METT-TC conditions the perimeter be broken into BPs of individual perimeter defense that are independent but interlocking for defense. The senior individual in each position is the commander for the perimeter defense. The ASB SOP covers as many defense procedures as possible.

5-155. Key support elements from the ASB are designated to evacuate the BSA to allow minimum support to the maneuver brigade should the enemy confront the BSA in sufficient strength to impact upon the ability to defend the BSA. The ASB should develop a displacement plan to support this requirement. However, all units must be able to defend against Level I activities (sniper, agents, saboteurs, or terrorist activities). They should be able to impede Level II attacks until assistance arrives. ASB units must defend themselves against attempts to disrupt their operations. They must be able to minimize destruction and reinforce their units. ASB units must also be able to gain time until response forces arrive.

5-156. If an enemy incursion exceeds the capability of response forces, TCFs must be committed to neutralize the threat. Assistance may come from a military police unit as a response force or TCF under control of the ASB commander. No logistics unit can sustain a defense against a determined Level II or III attack, but it should plan and train to protect itself until a TCF arrives to repel the enemy attack with assistance from the BSA. The ASB must be able to synchronize self-defense with BSA assets, military police, attached/OPCON maneuver units and the TCF when it arrives.

5-157. When the ASB commander plans in coordination with the aviation brigade S-3 for defense of the ASB's AO, he or she needs to have complete knowledge of—

- The elements in his sector of responsibility.
- The assets each unit has that allow it to defend itself.
- If the elements needed to defend against a large enemy threat are available.

5-158. Most supporting units (signal, engineer, and logistics) in the ASB's AO are located in the BSA. Sometimes due to METT-TC, many small elements form BPs with the entire group of BPs making up a BSA, which in itself is perimeter defense.

5-159. Commanders at all levels must consider, at some point, the time and effort used by sustainers to defend logistics locations degrades their ability to perform the support mission. There needs to be a dialogue between the aviation brigade commander and ASB commander regarding the ability of the BSA to conduct sustainment operations and its force protection requirements. There is a continuum of balancing requirements as the risk of enemy threat increases the amount of sustainment operations to be conducted decreases. The brigade commander and ASB commander must have this discussion as to what is a reasonable amount of risk to accept and then plan accordingly with as much risk mitigation as possible.

SECTION V – STANDARD ARMY MANAGEMENT INFORMATION SYSTEMS ARCHITECTURE

5-160. STAMIS (figure 5-4, page 5-28) consists of computer hardware and software systems that automate diverse functions based on validated customer requirements. STAMIS facilitates vertical and horizontal flow of logistics and maintenance status information to units Army wide.

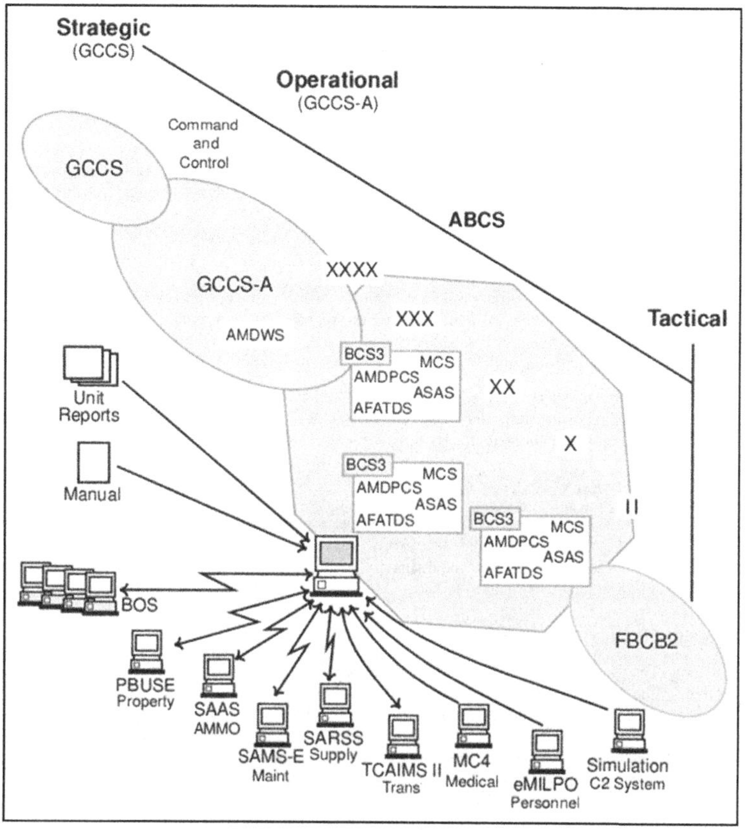

Figure 5-4. Standard Army management information systems architecture

STANDARD ARMY RETAIL SUPPLY SYSTEM

5-161. The standard Army retail supply system (SARSS) is a STAMIS for retail supply operations and management. The system includes all units and installations (regular Army, RC, and NG). SARSS provides supply-related data to the integrated logistics analysis program system.

5-162. SARSS comprises the following integrated systems:

- Standard Army retail supply system-level 1 (SARSS-1) at the SSA level.
- Standard Army retail supply system-level 2B at installation and U.S. Property and Fiscal Officer.
- Standard Army retail supply system-level 2A/C (SARSS-2A/C) at the corps and theater automated data processing service center.

- Standard Army retail supply system-gateway (SARSS-Gateway), formerly known as the objective supply capability.

UNIT-LEVEL LOGISTICS SYSTEMS

5-163. Unit level logistics systems (ULLSs) consist of software and hardware which automate the logistics systems for unit supply, maintenance, and materiel readiness management operations. ULLS prepares unit supply documents, maintenance management records, readiness reports, and property records. ULLS consists of three applications—ULLS-A, unit level logistics system-ground (ULLS-G), and unit level logistics system-supply (ULLS-S4).

UNIT-LEVEL LOGISTICS SYSTEM-AVIATION

5-164. ULLS-A enables aviation production control officers to generate and manage unit level work orders and post status to the maintenance request register. It also provides quality control officers automated component, inventory, and inspection master files. Production control receives a master maintenance data file updated and supplied from logistics support activity.

5-165. The Army materiel status system (AMSS) reporting capability within ULLS-A replaces the manual readiness reporting requirements outlined in AR 700-138. AMSS is the commander's link to monitoring the supply and maintenance posture of the unit.

UNIT-LEVEL LOGISTICS SYSTEM-GROUND

5-166. ULLS-G is located at units that have an organizational maintenance facility. It automates vehicle dispatching, PLL management, and the Army maintenance management system. The automotive information test interrogator is connected directly to the ULLS-G. ULLS-G is linked to the wholesale supply system through SARSS-Gateway.

UNIT-LEVEL LOGISTICS SYSTEM-SUPPLY

5-167. ULLS-S4 is located at unit-level supply rooms and battalion and brigade S-4 sections. ULLS-S4 automates the supply property requisitioning/document register process, hand/subhand receipts, component, budget, and logistics planning activities. It also receives and produces AMSS reports generated by ULLS-G, ULLS-A, or another ULLS-S4 system. The automotive information test interrogator is connected directly to ULLS-S4. ULLS-S4 interfaces with the standard property book system-redesign (SPBS-R), ULLS-G and ULLS-A (for budget and AMSS data transferring), standard Army ammunition system, STANDARD Army retail supply system-objective (SARSS-O) at the DS level, standard Army intermediate level logistics system supply, SARSS-Gateway, and BCS-3.

STANDARD ARMY MAINTENANCE SYSTEM

5-168. This system includes standard Army maintenance system-level 1 (SAMS-1) and SAMS-2.

STANDARD ARMY MAINTENANCE SYSTEM-1

5-169. SAMS-1 enables automated processing of DS/GS maintenance shop production functions, maintenance control work orders, and key supply functions. Requisitions are prepared automatically and automatic status is received from SARSS-1. SAMS-1 interfaces with other systems such as ULLS and SARSS-O. It also provides completed work order data to the logistics support activity for equipment performance and other analyses.

STANDARD ARMY MAINTENANCE SYSTEM-2

5-170. SAMS-2 is an automated maintenance management system used at the FSC and ASB level to—

- Enable monitoring of equipment not mission capable status, and control and coordinate maintenance actions and repair parts usage to maximize equipment availability.
- Receive and process maintenance data to meet information requirements of the manager and fulfill reporting requirements to customers, higher SAMS-2 sites, and the wholesale maintenance level. Data can be accessed instantly to enable management control, coordination, reports, analysis, and review.
- Provide maintenance and management information to each level of command from the user to the wholesale and DA levels.

STANDARD ARMY MAINTENANCE SYSTEM-ENHANCED

5-171. SAMS-enhanced integrates ULLS-G, SAMS-1 and SAMS-2 by incorporating the Windows graphical user interface operating systems. Standard Army maintenance system-enhanced (SAMS-E) acts as a bridge between current functionality and the enterprise resource planning solution. The following are benefits with SAMS-E:

- Fully replicates the functional capabilities of the current legacy systems—ULLS-G and SAMS-1 and 2.
- Reduces the number of computers and operators on the battlefield.
- Operates in the Windows 2003/XP environment that fully replicates the capabilities of the three legacy systems; ULLS-G and SAMS-1 become integrated and utilizes the same relational database as SAMS-2.
- Enables ordnance corps' two-level maintenance concept.

PROPERTY BOOK AND UNIT SUPPLY ENHANCED PROGRAM

5-172. Property book and unit supply enhanced provides close to real-time, accurate visibility of the unit's property book account operating on the AKO portal. The following are benefits with the use of property book and unit supply enhanced:

- Replaces two legacy systems—SPBS-R and ULLS-S4.
- Accurate visibility of unit level weapons systems and stocks.
- Use one common platform (light weight Pentium laptop) versus multiple platforms.
- Operational support with web-enabled capabilities (operates on any computer with web connection).
- Provides office automation, e-mail, on-line help and end user manual and automated catalog changes.
- Provides support for unit transfer/TF/split operations.
- Centralized database eliminates the need for thousands of smaller.
- Collaboration and interoperability provided by a common source of information required to support war planning via the global combat support system and joint C2 system.

INTEGRATED LOGISTICS ANALYSIS PROGRAM

5-173. The integrated logistics analysis program family of existing and planned management information utilities provides logistics and resource managers with integrated views of cross-functional data. Data are taken from the STAMIS at local, regional, and national levels, and Defense Finance and Accounting Service. These data are then integrated and displayed at levels of aggregation appropriate for each management level.

DEFENSE AUTOMATIC ADDRESSING SYSTEM

5-174. Logistics information processing system, which is maintained by the defense automatic addressing system, is DOD's central repository for information on the status of requisitions. It also augments global transportation network in monitoring the status of nonunit cargo shipments.

Appendix A
Ready, Deploy, and Redeploy

Any discussion of deployment and redeployment begins with an understanding of Army Force Generation (ARFORGEN). Once ARFORGEN is understood the discussion continues with an explanation of deployment; RSOI; and redeployment of personnel, ground vehicles, equipment, and aircraft. Aviation's requirements for ARFORGEN are to produce a trained and ready aviation force package; deploy and fight that force effectively; and once mission complete, safely and efficiently redeploy.

SECTION I – INTRODUCTION

A-1. The fundamental posture of the Army is power projection. For the Army to fulfill its role, it must be capable of rapidly deploying trained and ready force packages to any potential theater of operations and be able to achieve the military objectives set by the appropriate command.

A-2. Aviation units are among the first deployable package units and set conditions for follow-on forces. Aviation's unique ability to provide reconnaissance, security, CCA, ISR, C2 support, aeromedical evacuation, and sustainment support allows the sequencing of forces to accomplish the commander's intent by placing critical capabilities required in the AO first, and increasing the force tailorable package over time.

A-3. Aviation units publish detailed SOPs, conduct training, and develop plans to support deployment, RSOI, and redeployment. SOPs describe important preparatory activities such as personnel recall and preparation of aircraft, vehicles, and equipment for overseas shipment. In addition to conducting training in tasks related to post-deployment operations, aviation units also conduct training in tasks associated with deployment operations. Training occurs before and after aviation units are alerted for deployment. Likewise, planning for deployment and contingency operations begins prior to deployment notification. After notification for deployment, aviation units refine existing plans to account for new operational considerations not addressed in the original contingency plan.

SECTION II – ARMY FORCE GENERATION

A-4. ARFORGEN is a structured progression of increased unit readiness over time. This results in recurring periods of availability of trained, ready, and cohesive units prepared for operational deployment in support of regional combatant commander requirements. ARFORGEN is the process Forces Command uses to generate capable land forces for the JTF commander. The tenets of ARFORGEN are modular forces configured into tailored force packages, resourced and trained to the capability required for the specified mission, progressing in capability over time, with commanders responsible for determining their unit capabilities.

A-5. ARFORGEN defines how to organize, and provides the basis for how to equip, resource, train, and deploy forces. It is a progressive readiness strategy that builds on increasing capability levels. Combat capable forces are the critical output. The ARFORGEN standard for units is "ready for what = resourced for what = report against what." The "what" in this standard is the unit's sourced mission requirement (figure A-1, page A-2).

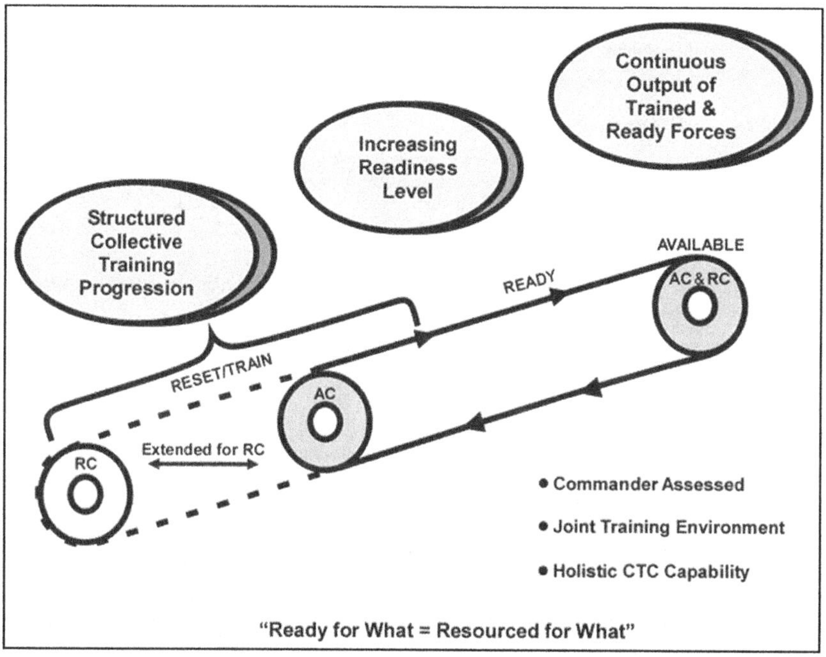

Figure A-1. Army force generation training and readiness strategy

A-6. ARFORGEN is a mix of force packaging and pools. The force packages provide mission focus for the development of METLs and focuses resource priorities across the force. Force pools (reset/train, ready, and available) are a management process by which the Army is able to provide flexible options for contingency planning and decisionmaking.

A-7. The ARFORGEN training and readiness strategy focuses on output, providing the requisite number of ready units required by the combatant commander. It is a progressive, gated, manning, equipping, and training strategy permitting commanders to constantly train to the highest level possible. It is driven by the force package mission, mission assumption dates, and METL. The strategy is designed to move units as quickly as possible through gates in order to develop operational depth and meet operational requirements. Training resources are prioritized from those nearest the fight to those farthest away. Commanders are not limited to floor or phase capability levels and may progress quickly to the highest capability levels achievable.

FORCE PACKAGES

A-8. ARFORGEN is a mix of three force packages—deployment expeditionary force (DEF), contingency expeditionary force (CEF), and ready expeditionary force (REF).

DEPLOYMENT EXPEDITIONARY FORCE

A-9. DEFs are task organized units designed to execute planned operational requirements and those currently executing deployed missions.

Note. RC units in a DEF are sourced against a future requirement, have been mobilized, or are currently mobilized.

CONTINGENCY EXPEDITIONARY FORCE

A-10. CEFs are the remaining available force pool units (not in a DEF) task organized to meet OPLANs and contingency requirements. These forces are capable of rapid deployment but are not yet alerted to deploy (regular Army) or mobilize (RC). CEF forces transition into DEFs if alerted.

READY EXPEDITIONARY FORCE

A-11. REFs are task organized units designed to train/prepare for potential future operational requirements or task organized to best execute full spectrum training.

FORCE POOLS

A-12. ARFORGEN provides a continuous output of capable forces focused on sourced operational requirements. It is the basis for allocation of resources to ensure forces are prepared to defend the homeland, focus on critical regions, swiftly defeat the opposition, and win decisively. Forces move in a cyclical progressive manner through the force pools progressing in capability to the available force. Generally, forces in reset/train concentrate on reconstitution and achieving organizational capabilities. The ready force contains units conducting mission preparation and higher level collective training and provides strategic depth to sustain on-going operations or meet unanticipated threats. The available force provides initial response regular Army and RC forces.

A-13. All units projected from CONUS are rotational force pool units and move through the three force pools discussed below to progress from reset/train, ready force to the available force based on sourced mission requirements which assign them to expeditionary force packages. Units are equipped at MTOE level to achieve capability and meet projected Army force requirements. Once sourced to a specific mission, equipment types and level are tailored to meet operational requirements.

RESET/TRAIN FORCE

A-14. A reset/train force includes units that redeploy from long term operations, are experiencing significant reorganization, or are unable to sustain ready or available force floor capability levels. The reset/train force begins upon either a unit's return from deployment or completion of 1 year in the available force. Units immediately begin reconstitution activities to bring equipment and personnel levels up to prescribed reset/train-day levels. Reset/train-day is a coordinated approved date, codified in an order and is the beginning of the unit's operational readiness cycle. DEF units perform theater focused training beginning on reset/train-day. CEF/REF units focus training on their core organizational mission.

A-15. Reset/train events and activities normally begin with a change of command, unit reconstitution to authorized unit strength levels, and conduct new/displaced equipment fielding and training. Units develop individual and crew gunnery, as required, and work toward staff proficiency based on core mission essential tasks. During the reset/train phase, the aviation brigade may be required to provide a platoon or company to a BCT for training and integration. Commanders move their units as quickly as possible into the ready force pool by attaining required capability levels. The gated event for progression to the Ready Phase is battalion level proficiency certified by a CTC, FTX, battalion level EXEVAL or equivalent.

READY FORCE

A-16. The ready force pool consists of ARFOR capable of performing at floor unit capability levels, conducting mission preparation, and executing higher level collective training with other horizontal and vertical operational headquarters. These units focus their training on their "ready for what" sourced METL.

RC forces in the ready force pool may be alerted, mobilized, and required to conduct post mobilization training for sourced operational missions. All units focus training on assigned DEF, CEF, or REF missions.

A-17. Ready force pool training focuses on assembling the aviation brigade package to sustain staff proficiency and attain ABTF level maneuver proficiency. During this phase the aviation brigade may execute a linked ATX within the higher headquarters mission readiness exercise or be validated by live participation in the mission readiness exercise. Headquarters and organic subordinate units participate in battle command training programs, combat training centers (CTCs), and joint exercises as required. Deploying units conduct theater specific mission focused training and participate in the mission readiness exercise. Additionally, the aviation brigade may provide subordinate units or ABTFs to support BCT CTC events. Regardless of whether a unit is designated DEF, CEF, or REF, they train to achieve proficiency on the respective theater or OPLAN mission(s). The aviation brigade is ready to move into the available force pool upon successful completion of support brigade warfighter exercise phase II embedded in MRX, BCTP, or standalone BCTP supported ATX. (ATX may be used as a separate gated event prior to MRX).

AVAILABLE FORCE

A-18. These units are capable of conducting theater specific or full spectrum operations. All regular Army and RC units begin a 1-year available force pool based on an assigned mission available date. DEFs are either prepared to deploy as indicated on execution orders, deploying, or deployed. The CEFs constitute the remaining available force pool units not deployed and are task organized to meet contingency plan and unanticipated requirements. These CEF forces are capable of rapid deployment and minimal pre-mission training. All CEFs become DEFs upon alert notification. A unit may receive an MTOE revision along the force modernization path prior to execution of the mission readiness exercise, but no further MTOE changes will be accepted after completion of the mission readiness exercise. Units return to reset/train after 1 year in the available force pool, upon returning from deployment, or on order from Forces Command.

A-19. Available force pool training for CEF units focuses on OPLAN mission requirements and sustaining full spectrum capabilities. The DEF units train on theater-related METL. The aviation brigade may be alerted for deployment as a brigade or ABTF.

SECTION III – DEPLOYMENT

A-20. Deployment is the movement of forces and materiel, and their sustainment, from point of origin to a specific AO to conduct operations outlined in a plan or order. It encompasses all activities from origin or home station through destination, specifically including intra-continental U.S., intertheater, and intra-theater movement legs. This combination of dynamic actions supports the combatant commander's concept of operations for employment of the force.

A-21. Deployments consist of the following four distinct and interrelated phases, which may not be sequential and could overlap or occur simultaneously:

- Predeployment activities.
- Movement to and activities at the POE—fort-to-port.
- Movement to port of debarkation (POD)—port-to-port.
- RSOI—port-to-destination.

A-22. Successful deployment planning requires knowledge of the unit's deployment responsibilities, understanding of the total deployment process, and intellectual appreciation of the link between deployment and employment. Deployment planning is an invariable, logical process focusing on Soldiers and equipment for deployment, ways to deploy them, and information and means to track them. In particular, deployment plans require specific, detailed information to guide the unit through an effective deployment. The heart of deployment planning is a precise list of Soldiers and equipment that must deploy—the unit deployment list, developed in the Transportation Coordinators' Automated Information for Movement System II. Its importance exemplifies its use; such as, manifesting unit equipment for deployment and updating the time-phased force and deployment data (TPFDD) so the appropriate lift is scheduled for the deployment.

A-23. Successful deployment planning requires knowledge of the unit's deployment responsibilities, an understanding of the total deployment process, and an intellectual appreciation of the link between deployment and employment. Steps used in planning and preparation during predeployment activities include—

- Analyze the mission.
- Structure forces.
- Refine deployement data.
- Prepare the force.
- Sechedule the movement.

FMI 3-35 provides additional information.

DEPLOYMENT TRAINING

A-24. Training allows aviation units to minimize post-notification predeployment activities. Deployment training addresses critical tasks related to movement facilitating aviation's ability to deploy. Although much of this training is focused on individual technical skills and applicable primarily at the company level, brigade and battalion staffs participate in deployment training for purposes of staff supervision and addressing deployment requirements.

UNIT MOVEMENT OFFICER TRAINING

A-25. Army policy requires company unit movement officers (UMOs) to plan and coordinate technical aspects of unit sea, rail, and air movement. The brigade and battalion staff is responsible for conducting deployment planning; however, it is useful for the staff to have some personnel—particularly in the S-3 and S-4 sections—trained in the technical details of associated duties. The scope of UMO technical training includes—

- USAF airlift operations.
- Characteristics and capabilities of the types of vessels, aircraft, or railcars the unit may use for deployment.
- Highway, rail, and port operations.

UNIT LOAD TEAM TRAINING

A-26. Aviation units must have appropriate numbers of personnel trained on aircraft and vehicle preparation and aircraft, sea vessel, and rail loading/unloading techniques. Brigade and battalion staffs, as well as companies, should designate load teams and train them in the following tasks:

- Preparation of aircraft and vehicle load plans.
- Preparation of vehicles for shipment by reducing operational dimensions, protecting fragile components such as windshields and mirrors, and weighing and marking procedures for air and rail modes.
- Tie-down procedures for vehicles, aircraft, and railcars.
- Operation of vehicles in conditions simulating loading/unloading techniques for aircraft and rail.

HAZARDOUS MATERIAL TRAINING

A-27. Each company, and the brigade and battalion S-4 sections, should have at least one school-trained person to certify hazardous cargo. Hazardous cargo certification teams visit most installations annually or semiannually. Regardless of whether the company has anyone who is school-trained or not, the UMO/NCO should be familiar with the contents of TM 38-250 and Code of Federal Regulations 49. Military Standard-129 and TC 38-3 contain additional packing assistance.

Standing Operating Procedure Development

A-28. Aviation units use SOPs to document critical deployment tasks beginning with alert activities, and continuing through RSOI to redeployment. Actions described in SOPs require periodic training and updating as deployment requirements change.

A-29. Examples of topics related to deployment that must be addressed in SOPs include—

- Family readiness group (FRG) functions and resourcing.
- Personnel recall procedures to include alert roster implementation responsibilities and recall of personnel on leave and temporary duty.
- C2 requirements.
- Local security.
- Media guidance.

Family Readiness Group Planning

A-30. FRG main goals are to—

- Serve as a link between the deployed unit and families, whether those families remain at home station or not. In this capacity, the FRG serves as a conduit for command information on deployment and redeployment dates, changes in the unit's status or mission, and other items of interest to family members.
- Facilitate deployment and redeployment briefings and activities in conjunction with the unit commander, rear detachment commander, and installation support agencies.
- Serve as a mutual support group for family members, stepping in with advice, personal counseling, or assistance when families have problems during unit deployment.

A-31. The most effective FRGs are those established as part of the unit's ongoing and routine mission preparation, rather than those created just prior to deployment.

A-32. The primary factor in determining the success of a unit's FRG is the energy levels of FRG leaders. Company- and battalion-level FRG operations live or die based on the energy of the personnel leading the family support initiatives at home station and on command emphasis placed on FRG operations by the unit's leadership. In most cases, the best approach to identifying leaders of the FRG is to ask spouses to volunteer. Successful FRG leaders generally are charismatic, people-oriented, caring people who tend to volunteer their time in the community in other ways as well.

Contingency Plan Refinement And Post Alert Training

A-33. Deploying aviation units access their higher headquarters' and the overseas gaining command's tactical intelligence home page using the All Source Analysis System (ASAS) to obtain information concerning threat, terrain, weather, and other data concerning the area in which the unit will deploy. This information, along with other operational guidance from higher headquarters is used to drive refinement of existing contingency plans.

A-34. Aviation units allocate time for mission-related training during deployment activities in response to changes and to develop additional proficiency in mission essential tasks. Use of virtual and constructive simulations (AVCATT, Longbow copilot trainer, synthetic flight training system, and ATXs) is efficient options for conducting this type of training.

ALERT ACTIVITIES

A-35. Notification for deployment may occur as part of an operation planned many months in advance, or it may happen as a short notice response to a crisis or emergency. In the case of little or no notice, aviation units must respond quickly and efficiently by mustering personnel, preparing aircraft, vehicles, and equipment for shipment, establishing security as required, and using any available time to conduct further predeployment training.

Personnel Considerations

A-36. The aviation unit's S-1 and S-1 section plan personnel support for deployment operations. This plan is based on the organization's deployment timeline; identification of appropriate times and locations for completion of personnel asset inventory medical screening; and preparation for overseas rotation events and other personnel activities such as—

- Identifying personnel shortages.
- Identifying nondeployable personnel.
- Initiating recall of personnel attending schools, on leave, and temporary duty.
- Requisitioning additional personnel.
- Identifying personnel records to deploy with the unit.

Preparation of Equipment for Deployment

A-37. As soon as sufficient personnel are mustered, or after movement to the aerial port of debarkation (APOD)/sea port of debarkation (SPOD), units should begin preparing aircraft, vehicles, and equipment for deployment. Brigade and battalion staff planners must be proactive in identifying requirements during coordination with supporting installation staff or transportation elements. Cargo and other equipment are prepared for shipment based on the SOP and instructions from supporting installation transportation authorities.

Command Post Establishment

A-38. Preparation for deployment requires effective C2. In addition to conducting internal C2 procedures, aviation units must use all available assets, such as ABCS and commercial service, to maintain contact with higher headquarters, supporting installation staff, and the gaining command. Shortly after alert or deployment notification, aviation organizations establish CPs to control various deployment activities. Initially a CP may be established as an emergency operations center, or other SOP-directed variant, to control recall activities and maintain communications. As elements begin movement and staging activities it may be useful to establish other CPs with functions similar to those of a TAC CP.

Family Readiness Group Considerations

A-39. A functional FRG has many tasks when a unit is notified of a deployment. Commanders and FRG leaders need to screen for Soldiers whose family members have special circumstances or special care needs. In addition to screening family members for pregnancies, exceptional family member medical conditions, and other situations prior to deployment, units should be aware of family members who might not speak English in order to plan alternate ways of keeping these family members informed. After identifying individuals with special circumstances, FRG personnel must adapt an overall plan to meet the needs of these families.

A-40. An FRG should also determine which spouses plan to depart home station after the unit deploys. Some spouses may want to live with relatives for support during the deployment, so units and FRG leaders need to make special provisions disseminating information to them, and assisting with medical and dental care and other personnel services. Specific problems may also include access to other military facilities.

Force Protection Considerations

A-41. Planners must give careful consideration to all possible threats during deployment operations and then implement appropriate levels of OPSEC and force protection. In some cases it may be necessary to limit access to certain areas in the aviation unit's garrison offices or impose traffic and parking controls. Security during a road movement to an APOD/SPOD requires coordination with supporting installation or higher headquarters staff for local law enforcement agency support and assistance.

Media Considerations

A-42. Media impact on deployment operations is substantially greater today than any previous time in history. The news media's capability to gain and transmit ongoing deployment activities globally must not be discounted. News technology requires establishment of a single POC for releasing information regarding ongoing operations. The aviation brigade's PAO coordinates all media actions and responds to public requests for information as appropriate. The higher headquarters develops procedures and guidelines for releasing information within security, propriety, and safety considerations of the ongoing operation.

Establishment of Rear Detachment

A-43. In some cases a rear detachment may be designated when an aviation element deploys. The rear detachment commander and personnel have responsibilities for all personnel and equipment remaining at home station during a unit's deployment. Additionally, the rear detachment has significant family readiness responsibilities and provides FRG leaders with a point of entry into official Army financial, legal, and other personnel services systems. A positive and supportive relationship between rear detachment personnel, the deployed unit, and FRG leaders is critical to the overall success of the family readiness effort.

METHODS OF DEPLOYMENT

A-44. Aviation normally deploys in one of two methods—sea and air transport, or self-deployment.

Sea and Air Transport

Planning and Preparation

A-45. Successful movement depends on detailed planning, SOPs for deployment by various methods, and identification, training, and validation of deployment and load teams. Each team member has specific duties, from preparation at home station, to clearance of the POD, to arrival at destination. The unit must continually validate automated unit equipment lists and time phase deployment lists in preparation for future deployments.

A-46. Upon receiving the WARNO, and time permitting, advance parties are sent to POEs and PODs to set conditions for reception of unit personnel and equipment, and provide command, control, and communications.

A-47. The following references discuss deployment actions and considerations:

- UMO Deployment Handbook Reference 97-1.
- FM 3-04.500.
- FM 4-01.41.
- FM 4-01.30.
- FMI 3-35.
- CAB TACSOP.

A-48. Surface Deployment and Distribution Command Transportation Engineering Agency (SDDCTEA) pamphlets provide specific guidance for preparation of equipment for movement. The following pamphlets can be downloaded from http://www.tea.army.mil/pubs/deploy.asp:

- SDDCTEA Pamphlet 55-19.
- SDDCTEA Pamphlet 55-20.
- SDDCTEA Pamphlet 55-21.
- SDDCTEA Pamphlet 55-22.
- SDDCTEA Pamphlet 55-23.
- SDDCTEA Pamphlet 55-24.
- SDDCTEA 70-1.

- SDDCTEA Pamphlet 700-2.
- SDDCTEA Pamphlet 700-4.
- SDDCTEA 700-5.
- SDDCTEA Pamphlet 700-6.

A-49. Aircraft preparation, lifting, and tie-down must be in accordance with appropriate preparation for shipment manuals and specific loading manuals for military aircraft (FW air shipments only). The following TMs can be downloaded from https://www.logsa.army.mil/etms/online.cfm:

- EM 0126. TM 1-1520-Apache/Longbow.
- TM 1-1520-237-S.
- TM 55-1520-238-S.
- TM 1-1520-248-S.
- TM 1-1520-241-S.
- TM 1-1520-252-S.

A-50. Not all contingencies for unit movement can be foreseen due to the wide range of missions and world events that may occur. Units must be aware of battle book plans, and wargame probable and possible scenarios. Skeleton plans are established to cover contingencies.

A-51. Unit movement personnel are familiar with the POEs available to their organization and mission requirements. Special needs and considerations are addressed as early as possible for each POE. Unit movement personnel—

- Establish and periodically update telephone lists, points of contact, and special requirements for likely POEs.
- Conduct periodic leader reconnaissance of POEs to include members of unit load teams and advance party personnel.
- Identify advance party personnel and define duties.
- Identify OPSEC requirements during movement and embarkation activities.
- Plan and coordinate workspace for personnel during the embarkation phase (empty offices, borrowed tentage from nondeploying units, and rented or borrowed trailers).
- Identify and prepare requests for communications requirements (commercial lines, wire, radio, and cellular phone).
- Determine transportation requirements at POE for movement teams and key personnel (borrowed vehicles and rental cars).
- Plan messing, billeting, MTFs, refueling/defueling points, and special requirements for weapons and ammunition.

Movement

A-52. Upon receiving the order, units ferry their aircraft and move ground vehicles along preselected routes to the POE. Units performing depot-level maintenance normally operate at these embarkation points. As the units arrive, a dedicated sustainment support team assists in preparing vehicles, equipment, and aircraft for deployment. Preparation includes required maintenance and installation of ferry equipment.

A-53. On receipt of the deployment order, ASB commanders dispatch preselected facility teams. Deployment headquarters staff members locate command facilities at each termination site to facilitate integration of aircraft, vehicles, and personnel into the theater force structure.

SELF-DEPLOYMENT

Planning Considerations

A-54. Self-deployment is an alternative method used to rapidly move aircraft. Units consider the following factors when planning self-deployments:

- Securing departure, flight routes, and arrival points prior to movement.
- Establishing proper facilities, personnel, equipment, and supplies at the destination ensuring quick transition to operations.
- Pre-positioning ground support teams at stopover points along self-deployment flight routes. Ground support teams include personnel, equipment, and repair parts to provide limited services (POL products, supply, ammunition, HSS, communications, weather forecasting, and flight planning).
- Avoiding self-deployment over large bodies of water except in an emergency when other methods are not available. Overwater operations require extensive CRM; crewmembers must be trained and proficient with specialized ALSE and emergency procedures for overwater flight. For extensive legs of overwater flight, plans should include naval assets along the flight route to provide intermediate fuel stops or SAR.
- Minimizing deploying combat troops on self-deploying aircraft. Available space is typically used to accommodate those supplies, tools, parts, survival equipment, and limited support personnel necessary to make flights self-sustaining during the deployment.
- Coordinating alternative transport of some weapon systems, equipment, and baggage for aircraft self-deploying over long distances.
- Ensuring ASE is properly functioning and calibrated (TACOPS provides proper codes for AO), even when traveling over assumed friendly territory.
- Coordinating attack reconnaissance aircraft (or any aircraft without cargo capability) to travel with lift assets carrying emergency supplies, maintenance equipment and personnel, and backup aircrews in case of an unscheduled landing.
- Ensuring aircrews are familiar with established downed aircrew and aircraft recovery procedures.

Personnel

A-55. Aircrews and passengers may require passports and visas for each country of intended landing. The mission may require crew members or other support personnel with specific foreign language proficiency for those countries in which refueling or extended stopovers are planned.

A-56. Extensive distances may require aircrews to fly many hours. The challenge is ensuring crews are able to conduct operational missions upon arrival in-theater. Commanders adjust work and rest schedules before and during deployment. Commanders must plan to rotate crews through pilot duties whenever possible. Deploying units could carry backup crews from nondeploying units on CH-47 and UH-60 aircraft.

Intelligence

A-57. Units obtain threat intelligence information for those countries that are overflown and where landings are planned. Terrorist threats, counterintelligence, and specific force protection concerns are important to aircrews for planned and potential stops.

A-58. Routes into possible hostile airspace should be avoided. If unavoidable, unarmed aircraft must be escorted.

Training

A-59. Commanders must place emphasis on predeployment training. This training includes water survival, ALSE functions, fuel system management, high gross-weight operations, International Civil Aviation Organization flight planning, navigation equipment, communication requirements, shipboard operations, and rescue operations.

A-60. En route and destination environmental considerations—such as high altitude, mountainous and jungle terrain, and overwater flight—are considered. Crews must be trained for survival in the environment and use of special equipment required for each environment.

A-61. Theater-specific ROE, Status of Forces Agreements, local customs, language training, and OPSEC requirements that can be anticipated should be performed at home station, if possible.

Logistics

A-62. Self-deploying and supporting units request and coordinate maintenance and crew-rest facilities, fuel, transportation, security, and messing for stopover-point teams and self-deploying aircrews. If U.S. ground support teams are not available, units coordinate with friendly nations to provide required services. The S-9 acts as the POC for staff officers dealing with host nations. If no S-9 is assigned, the S-3 performs this function.

A-63. When aviation units deploy to destinations lacking fixed-base facilities, pre-positioned ground support teams perform those functions. S-4s of self-deploying and supporting units are responsible for logistics requirements along the self-deployment route and at the destination. Aviation MOs organize a maintenance support operation to prepare aircraft for self-deployment and meet maintenance requirements along the route and upon arrival at the destination.

A-64. Staff members verify the availability, quantity, and type of fuel at en route fuel stops, rather than depend solely on Department of Defense flight information publications (DOD FLIP). An appropriate agency verifies fuel quality at each location before refueling.

A-65. If required at stopover sites, contracting officers or Class A agents should be members of the advance party.

A-66. Units issue appropriate survival equipment and clothing for climates encountered in the route of flight.

A-67. To facilitate mission readiness movement planners, logisticians, and maintenance personnel carefully wargame arrival of units and equipment into the theater.

Mission Planning

A-68. AD identification zone procedures, as well as international interception signals, must be clearly understood by all aircrew members.

A-69. If applicable, all aircrew members must obtain and understand approved international clearances before departure. The SP and TACOPS officer provide assistance in disseminating the SPINS/ACO/aviation procedures guide.

Flight Organization and Aircraft Configuration

A-70. Each departing flight must contain multiple aircraft and be self-sustaining in terms of food, water, limited maintenance capability, and force protection. Aircraft with limited cargo capacity—such as AH-64s or OH-58Ds—require task organization with UH-60s or CH-47s. USAF or naval support is essential for downed aircrew recovery. Ideally, an escort PR aircraft is assigned. Without escort, each flight should include at least two aircraft with rescue hoists.

A-71. MPs and personnel are included in each flight and/or are pre-positioned at various planned stopover locations.

A-72. Depending on the type of aircraft and space available, a maintenance support package might include an auxiliary fuel system, tow bars, packaged POL, limited spare parts, a mechanic's toolbox, and tug or tow vehicle.

Communications

A-73. Units must—

- Coordinate frequencies for internal flight following throughout movement.
- Coordinate and verify compatibility of specific frequencies for supporting naval vessels and PR elements.

- Take SATCOM sets if available; SATCOM enables each flight to communicate its status to home station and theater of operations.
- Coordinate and deconflict frequencies for overflight of international/host nation airspace.
- Equip advance parties with communication equipment and flight frequencies for arrival at stopover points and the POD.
- Coordinate and verify compatibility of specific frequencies for JSTARS or airborne warning and control system (AWACS) monitoring of aircraft movement within a particular theater of operations.
- Coordinate for proper SPINS/ACO.

Equipment

A-74. Survival vests, rafts, survival kits, rescue hoists, survival radios, food, and water are essential mission equipment. Units maintain a critical equipment list at home station for planning considerations.

A-75. Each flight should have multiple aircraft with extra survival equipment that can be dropped to downed crewmembers.

Weapons

A-76. Individual and crew-served weapons should normally remain out of sight during flight and ground operations. They should be loaded, but not armed, to assist with protection from possible sudden engagements.

A-77. The controlling headquarters issues ROE when deploying units carry weapons and ammunition.

SECTION IV – RECEPTION, STAGING, ONWARD MOVEMENT, AND INTEGRATION

A-78. The goal of aviation RSOI planning to support efficient and timely equipment preparation at the POD and rapid deployment to an AA as required by the gaining higher headquarters. Communications and digital connectivity are established with subordinate, adjacent, and higher headquarters. Aviation RSOI activities are conducted concurrently with follow-on mission planning and preparation, and while maintaining required levels of force protection. The four RSOI steps are reception, staging, onward movement and integration.

- **Reception**: The process of unloading personnel and materiel from strategic transport; marshalling deploying units; transporting them to staging areas, if required; and providing life support to deploying personnel.
- **Staging:** The process of assembling, holding, and organizing arriving personnel and equipment into units and forces, incrementally building combat power and preparing units for onward movement, and providing life support for personnel until the unit becomes self-sustaining.
- **Onward movement:** The process of moving units and accompanying material from reception facilities and staging areas to tactical AAs or other destinations, moving arriving nonunit personnel to gaining commands, and moving arriving sustainment material from reception facilities to distribution sites.
- **Integration:** The synchronized transfer of authority (TOA) over units and forces to a designated component or functional commander for employment in the theater of operations.

A-79. RSOI generates combat power. Limited time requires that aviation units execute RSOI tasks rapidly to facilitate integration of essential aviation support into JTF operations.

PLANNING CONSIDERATIONS

A-80. Air and sea deployment modes terminate at the designated POD. Depot or ASB facilities are in the theater. Personnel at these facilities assist the unit with removal of ferry equipment, installation of mission equipment, and perform required maintenance and inspections to prepare equipment for the mission. They also coordinate immediate backhaul of designated support teams and ferry equipment.

ADVANCE PARTY ACTIVITIES

A-81. Advance party tasks are critical to the aviation unit's success during RSOI operations. Tasks the advance party may perform include—

- Rapidly establish CPs to control and coordinate RSOI efforts.
- Coordinate POD requirements with higher headquarters and supporting transportation organizations.
- Maintain communications with parent unit while the aviation unit main body deploys.
- Conduct coordination with gaining command to define future combat operations.

A-82. By performing these tasks the advance party establishes conditions to facilitate aviation integration into JTF operations.

TASK ORGANIZATION

A-83. Arriving elements task organize and reconfigure vehicles and aircraft as appropriate for the mission. Sustainment efforts are prioritized to build combat-capable units and C2 architecture.

FORCE PROTECTION

A-84. Aviation forces are particularly vulnerable during the build-up phase when the unit is not at full strength, and aircraft and vehicles may not be fully assembled for combat. The security plan must be understood and executed; the first priority of work immediately upon arrival at designated POD. This plan should include passive and active measures to combat air and ground threats.

A-85. Aviation forces are often among the first units to arrive in-theater. They may have to provide reconnaissance, security, and attack operations securing a lodgment before more forces arrive in-theater. This situation may require that aviation units conduct immediate and continuous operations from offshore or remote locations while the main body moves into the lodgment area.

A-86. To reduce risk of fratricide, crew members must understand—

- The ground maneuver plan.
- The commander's intent.
- The composition and location of friendly forces.
- Theater-specific IFF procedures.

TRAINING

A-87. Local area orientations, test flights, or other requirements not executed in advance may be required. Commanders should attempt to phase the arrival of personnel—such as mission training plans, test pilots, and key leaders—to begin before the unit's main body arrives. If units are already present in country, these key personnel deploy as early as possible to train with those units. The advance party is briefed on the requirements and plan for execution in order to identify and coordinate required external support.

A-88. Acclimation training may be required. Many units moving from one environmental extreme to another need to adjust to the new climate. The unit commander arranges training and conditioning to accelerate acclimation.

A-89. Most deployments involve operating in a joint or multinational environment. Early-arriving units may be able to schedule training with other services. Liaison elements from the S-3 shop are designated to ensure smooth coordination.

RELIEF-IN-PLACE

A-90. Upon arrival into theater, the unit may conduct a relief-in-place with another aviation unit. The relief-in-place is planned, coordinated, and executed by the relieved unit. This unit should coordinate with the unit being replaced to ensure all personnel are briefed on relief-in-place procedures prior to deployment into theater. The relief-in-place begins with arrival of the relieving unit's personnel and equipment and concludes with a TOA. The TOA signifies that the relieving unit possesses OPCON and mission execution requirements of the designated AO.

A-91. Relief-in-place procedures differentiate between each AO and type of unit replaced. However, aviation utilizes common planning factors common to most AOs. Each brigade, battalion, and company conducts relief-in-place tasks to prepare for TOA.

Tasks

A-92. The aviation brigade is responsible overall for ensuring all relief-in-place tasks are completed by battalions, companies, and attached elements. Common relief-in-place tasks completed by staff elements may include—

- Theater/country/AO briefs.
- Local area orientations.
- Leave/pass policies (procedures, locations, and emergency leaves).
- Brief/report formats, suspenses, and contingencies.
- Mail procedures.
- Officer/NCO evaluation report procedures (submission requirements for theater).
- Threat area and enemy weapons systems brief.
- Arms room, sensitive item military van (container), and courier card procedures.
- Battle update briefing procedures.
- Battle rhythm/CP shift manning procedures.
- Flight schedule, flying-hour report, and very important person procedures.
- Force protection.
- FS targeting and fire procedures.
- Falcon view, data transfer, airspace, and boundary crossing procedures.
- Range facilities (requesting, occupying, test fire, and capabilities).
- Classes of supply (requisition, storage, contract, and delivery).
- Property book procedures.
- Contracting officer training.
- Local nation vendor transactions.
- Life support systems orientation (laundry, gym, and morale, welfare, and recreation).
- Communications structure, operation, and accountability (secret internet protocol router [SIPR], nonsecure internet protocol router [NIPR], and retransmission).
- Pre-accident plan.
- Adjacent unit orientation/visit.
- TOA ceremony.

Battalion and Company Tasks

A-93. Common relief-in-place training tasks completed by battalion and company aircrew and personnel may also include—

- Mission orientation/execution (left/right seat rides).
- Emergency procedures (battle drills, mass casualty, and breach of security perimeter) training.
- Local area orientations.

  - Flight planning systems.
  - Flight operations and procedures such as flight planning area, ATO/ACO, SPINS, notices to airmen, ATS updates, helicopter LZ diagrams, flight mission boards and flight plans, weather facilities, weather briefings and updates, wire hazards data, maps, chart updates, DOD FLIP, and aircrew reading file.
  - Range test fire procedures.
  - High intensity radio transmission area information.
  - Boundary crossing procedures.
  - Local flying area.
  - ALSE procedures.
  - Aeromedical evacuation procedures (local medical facilities, frequencies, and telephone numbers).
  - Airfield layout and procedures (NAVAIDs, high intensity radio transmission area sites, arrival and departure procedures, hazards to flight, no fly areas, helipads and runways, instrument approaches, ATS information, and maintenance test flight areas).
  - Flight TTP.
  - Weather minimums.
  - Weapon control status.
  - Test fire procedures.
  - ROE.
  - Inadvertent IMCs.
  - FARP locations and procedures.
  - PR operations.
  - Aviation QRF operations.
  - Aviation procedures guide impact (airspace overview. altitude policies, lighting policies.
- Joint property inventories.
- Transfer of stay behind equipment.
- Installation property book transfer (barracks; automation; office equipment; communication equipment; and morale, welfare, and recreation).
- Continuity book handover.
- OPSEC program and procedures.
- Contingency plan review and handover.
- Joint logistics convoys and mission rehearsals/mission packets.
- Training and range resource overview.
- DART operations.
- Familiarization training for new equipment.
- Force protection program and procedures (guard mount, gate guard, and OP).
- Intermediate and DS maintenance operations.
- Fuel system supply point operations.
- SSA procedures.

SECTION V – REDEPLOYMENT

A-94. Aviation units conduct redeployment operations in order to meet the schedule established by higher headquarters, possibly in response to another contingency mission. When redeploying to home station in the CONUS, aviation units conform to the requirements of the U.S. Customs Service and the U.S. Department of Agriculture regarding shipment and condition of aircraft, vehicles, equipment, and cargo. Aviation units conduct redeployment while maintaining required levels of force protection.

A-95. Redeployment is not a stand alone operation, but a combination of continued daily tactical missions, a relief-in-place, and a plethora of redeployment activity. Units may use the R4 redeployment model during this operation. The four phases of the R4 model are redeploy, reintegration, reconstitution, and retraining (figure A-5) (see TC 1-400).

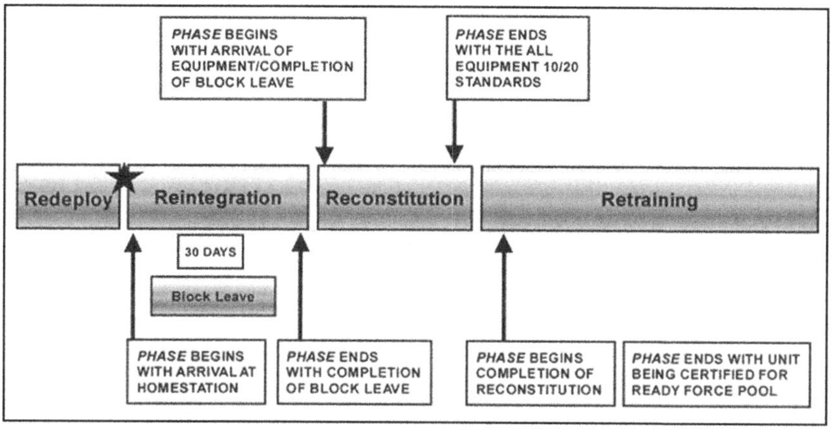

Figure A-2. Redeployment, reintegration, reconstitution, and retraining model

REDEPLOY

A-96. Aviation units may redeploy to home station or a different theater of operations. Aviation units redeploy in the following four phases:

- Recovery, reconstitution, and preparation for redeployment activities.
- Movement to and activities at POEs.
- Movement to PODs.
- RSOI when deploying to another AO.

A-97. Redeployment begins as directed and as METT-TC allows. It presents the same challenges to commanders with task organizing and echeloning forces as deployment. Preparation for redeployment begins as units start assisting other organizations with relief-in-place. Brigade and battalion staffs plan for redeployment using the same planning considerations during deployment operations as discussed in section I.

A-98. Initially, aviation units redeploy advance parties, and less essential personnel and equipment. During this stage, unit strength and equipment status may change often; however, accurate accountability is key to reconstitution. Additionally, commander's conduct training programs to maintain individual and unit METL proficiency. Aside from individual training tasks, units may conduct individual and crew served weapons qualification and familiarization, rehearse convoy operations, validate and update deployment SOPs, and conduct limited convoy live fire exercises (if time and resources are available). Protection of the force remains critical. If critical deployment personnel (UMO, hazardous materials, or logistics personnel) were lost during the operation due to expiration term of service, inter theater moves, or combat loss, commanders must coordinate training of additional personnel to perform these functions.

A-99. Aviation companies must perform vigilant maintenance on all equipment, especially aircraft and rolling stock. UMOs update and validate load plans, including input to the TPFDD. Units identify nonessential equipment and personnel for early return and identify stay behind equipment for follow-on units. Aviation units must also ensure LNOs and C2 cells are placed in key nodes and LNOs are knowledgeable of unit equipment levels and deployment procedures.

A-100. After an aviation unit is alerted for redeployment, arrangements should be made through the home station FRG to conduct family reunion briefings. It may be appropriate to provide counseling and briefings on likely sources of friction among family members after a lengthy separation for all deployed Soldiers prior to arrival at home station.

Rear Detachment Operations

A-101. Aviation units use a rear detachment during redeployment operations to conduct important tasks after the redeployment of the main body. Critical rear detachment tasks include—

- Preparing aircraft, vehicles, and equipment for redeployment.
- Conducting air, rail, and ship-loading activities.
- Maintaining communications with the main body after redeployment.
- Coordinating with U.S. Department of Agriculture personnel, U.S. Customs Service authorities, and supporting transportation and port authorities.
- Coordinating disposition of supplies and equipment that cannot redeploy.
- Coordinating turnover of facilities to host nation authorities or a relieving unit.
- Providing operational and threat update to relieving unit.

REINTEGRATION

A-102. The reintegration phase begins with arrival at home station and ends with completion of block leave. The focus of the reintegration phase is on getting the Soldier well.

A-103. The rear detachment plays a vital role during the reintegration phase. Critical tasks performed by the rear detachment may include—

- Receiving aircraft, equipment, and personnel.
- Coordinating redeployment briefings for arriving personnel.
- Coordinating with FRG for reception of personnel.
- Maintaining C2 until completion of block leave.

RECONSTITUTION

A-104. Reconstitution begins with arrival of the unit's equipment and completion of block leave, and ends with all equipment in reset or at 10/20 standards. Reconstitution places the unit in the reset/train force pool of ARFORGEN. During reconstitution, unit focus is maintenance intensive, as well as reestablishing C2 systems and personnel levels. Section II provides more information on reset/train force pool.

RETRAINING

A-105. Retraining begins with completion of reconstitution and ends with the unit being certified for the ready force pool. The retraining phase focuses on battalion and company METL, individual/crew/collective training, aerial gunnery, ASE/EW standard exercises, individual/crew small-arms qualification, and staff operations. Section II provides more information on reset/train force pool.

Appendix B
Communications

This appendix outlines communication tools and generalized TTP to ensure effective C2 of aviation, ground, and flight operations and current digital communications equipment designed to facilitate SA and SU. Communications personnel and equipment are one of the commander's most valuable assets, a fact not lost on planning cells targeting communications sites and nodes early in an operational campaign.

SECTION I – NETWORK SUPPORT COMPANY

B-1. Aviation brigade elements frequently operate over long distances, wide fronts, and extended depths from their controlling headquarters. Communications must be redundant and long range to meet internal and external requirements. Long range communications can be augmented through the aviation brigade's organic NSC. Emphasis on information sharing and network layers makes the NSC mission more important than before.

B-2. The NSC provides network and signal support to the aviation brigade headquarters as well as implementation of the communications plan for the entire aviation brigade with technical oversight provided by the aviation brigade S-6. The entire information system, to include the communications network, network regulation, and all the systems that process, store, and distribute information are installed, operated, and maintained by the NSC.

B-3. The NSC implements multiple layers of communications including terrestrial, airborne, and space layers (SATCOM) both commercial and military. The NSC provides communications implementation for the multilayered network and installs and maintains the links that establish the aviation brigade network with higher level and subordinate networks. The NSC also provides connectivity to Army external elements or joint/multinational elements via a link from the current generation of tactical common user systems. The NSC services subordinates installing, operating, and maintaining the TI that uses FBCB2 to provide SU and battle command data exchange to mobile users such as aircraft and tactical vehicles.

B-4. Within the aviation brigade CP, the NSC is responsible for providing connectivity for an entire spectrum of communications needs to include VTC capability via mobile subscriber equipment (MSE)/joint network node (JNN) or satellite VTC.

B-5. Part of the NSC's responsibility is to ensure communication means are available to carry the data the aviation brigade needs to perform its mission. The complex digital communications systems require proper connectivity and functional integration throughout the entire digital architecture before the aviation brigade commander can digitally communicate with confidence. A step-by-step check of individual and collective functioning of the ABCS is required to validate the architecture and troubleshoot the system.

B-6. The S-6, in conjunction with the NSC, verifies the digital systems architecture to ensure there is a plan to communicate with units in the task organization and higher headquarters. Since the internet protocol based unit addressing system does not allow for dynamic changes of units entering and leaving the task organization, the importance of digital architecture validation is the foundation for success. This architecture validation is a system-by-system check done in conjunction with the users' platform-by-platform (vehicles) check to ensure each individual system has all the required component parts and functions properly.

B-7. After architecture validation, connectivity testing of the upper and lower TI begins in each battlefield functional area. Every identified problem that is resolved must be retested to ensure the fix meets architecture standards which ensure connectivity and stability. A fix plan must be developed to ensure problems are solved in a fashion that strengthens the digital chain.

SECTION II – NETWORK ARCHITECTURE AND TACTICAL INTERNET

B-8. Network-centric warfare enables the Army and other U.S. forces to achieve an advantage through improved information sharing. The ability to develop and leverage this information advantage and use it to achieve increased combat power is key to success of the aviation brigade. Networking the force into a single virtual infosphere provides the warfighter with a distinct information advantage.

B-9. The Army's network architecture is the warfighter information network-tactical (WIN-T). It is an evolving tactical telecommunications system consisting of infrastructure and network components from the combined arms battalion to the theater AO. It comprises multiple systems and pathways designed to facilitate information distribution and access to information services.

B-10. The TI is one element of WIN-T. The TI consists of tactical communications radios, linked by routers, using commercial standards for addressing and information protocols. It allows digital systems to send and receive COP information and C2 messaging.

B-11. Both COP information and C2 pass simultaneously over the TI. The TI's design provides capabilities that are mobile, secure, survivable, seamless, and capable of supporting multimedia tactical information systems. These capabilities continue despite masking terrain, distance, enemy EW, loss of key signal elements or CPs, or replacement of individual platforms. The TI consists of two segments:

- A lower TI connecting echelons brigade and below.
- An upper TI providing inter-brigade and division connections and above.

LOWER TACTICAL INTERNET COMMUNICATIONS

B-12. Digital communications connectivity for COP information and other lower TI C2 data for brigade and below has three primary components:

- EPLRS—data-only communication (platform position and network coordination).
- Single-channel ground and airborne radio system (SINCGARS)—voice and data communications.
- Internet controller (INC)—routing and interface capability.

B-13. Not all FBCB2 ground or aviation platforms are EPLRS equipped. The non-EPLRS platforms pass FBCB2 data, via the INC, to servers with SINCGARS and EPLRS. Every platform is associated with an EPLRS server through which all COP information and C2 data are routed. Platforms consistently evaluate server quality and jump to an alternate server if the primary server output degrades. Vehicles and aircraft without FBCB2 require verbal reporting and manual tracking. Vehicles and aircraft with EPLRS and FBCB2 will have displays showing the COP, tailorable to the needs of that platform (figure B-1, page B-3).

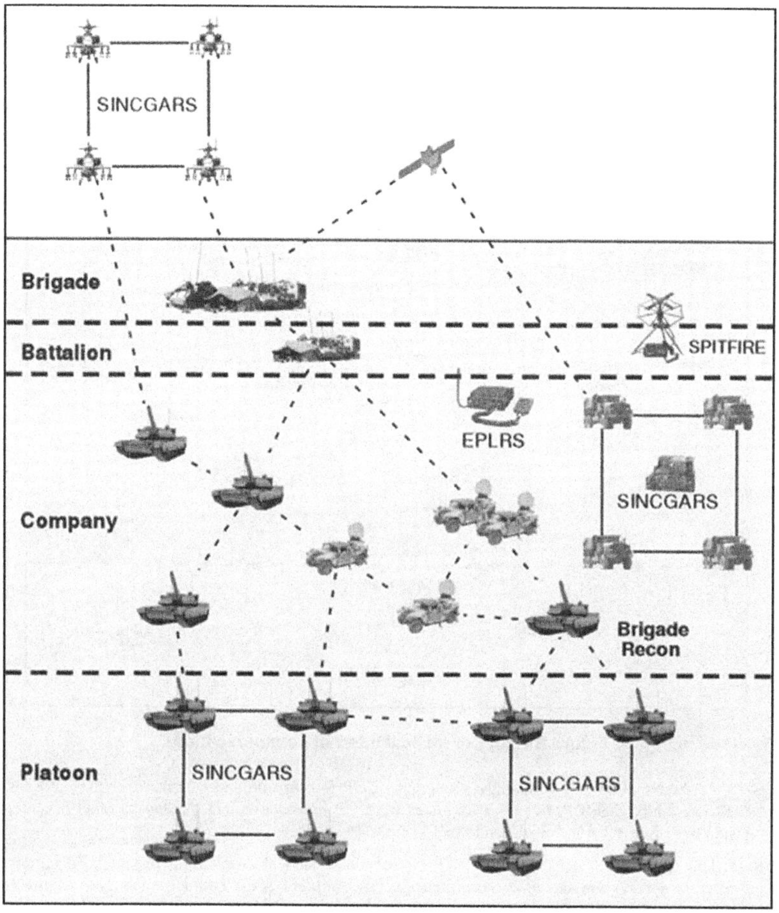

Figure B-1. Lower tactical internet communications

UPPER TACTICAL INTERNET COMMUNICATIONS

B-14. MSE and near term digital radio (NTDR) provide upper TI access/interface to battalion and brigade CPs. Upward dissemination of the FBCB2 COP and C2 data occurs over the upper TI between the battalion and brigade CP and to higher echelons. The upper TI also permits access to intelligence data of the Army battle C2 system of higher and adjacent headquarters. The NTDR handles the bulk of data between the battalion and brigade CPs with dissemination to and from higher headquarters via MSE (figure B-2, page B-4).

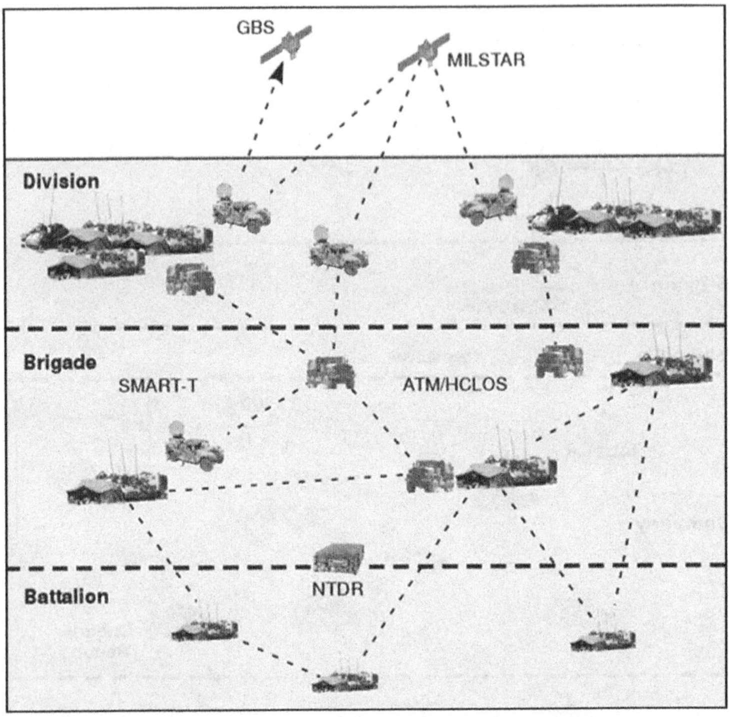

Figure B-2. Upper tactical internet communications

B-15. At the battalion CP, the CP INC routes the COP information and C2 data over the CP LAN and NTDR. At the brigade CP, the CP INC routes the COP information and C2 data to other brigades and division over the CP LAN, NTDR, and MSE.

B-16. The upper TI is part of the WIN-T that connects tactical echelons to distant headquarters and information sources via the global broadcast service, military SATCOM terminals and satellites, and high-capacity LOS transmission.

SEAMLESS TACTICAL INTERNET AND NETWORK ARCHITECTURE CONNECTIVITY

B-17. Both the upper and lower TI permits seamless exchange of COP information and C2 data (figure B-3, page B-5). Brigade is where such transfer occurs between the upper and lower TI. The following four elements are essential to sharing COP information and C2 messaging all the way down to a vehicle or aircraft platform:

- Appropriate radio waveform (SINCGARS, EPLRS, NTDR, and JTRS).
- Application software (FBCB2 and ABCS).
- Network architecture (TI and WIN-T elements).
- Platform processing and display (COP information software and display).

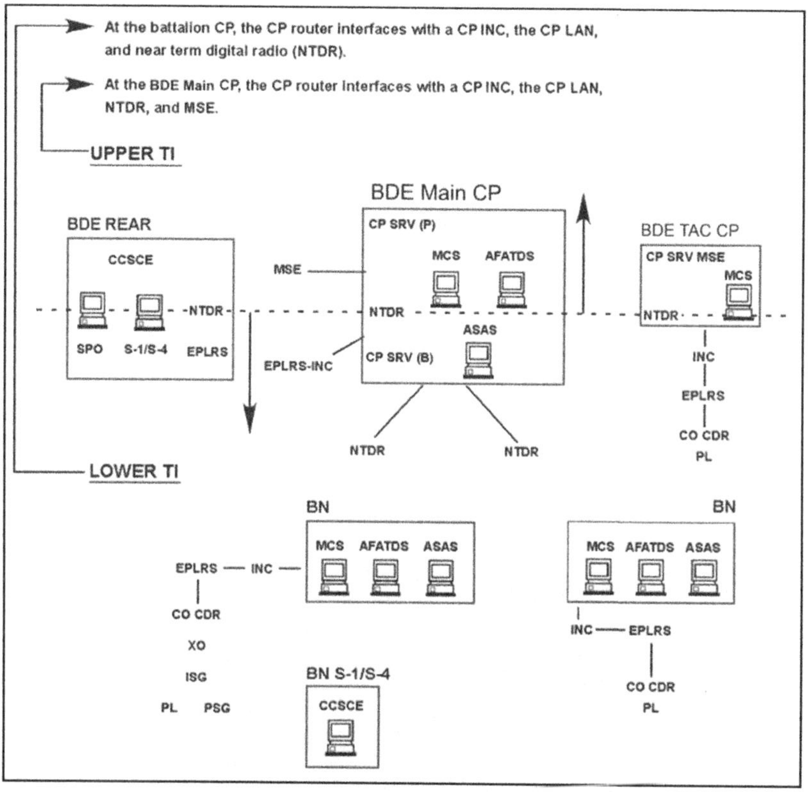

Figure B-3. Upper and lower tactical internet interface

JOINT NETWORK TACTICAL CAPABILITY SYSTEM

B-18. The Army's Joint Network Tactical Capability System is made up of two major components—the JNN found at brigade and higher levels and battalion level CPs module allocated to battalion level CPs. The joint network tactical node (JNTC) architecture has three main components—hub node, JNN, and battalion level node.

B-19. The hub node is designed to support division-level forces. The JNN, located at brigade/BCT level, connects into the hub using satellite links. However, the hub node could deploy as GS to a brigade or BCT depending on mission circumstances.

B-20. The division hub, which could be GS to the aviation brigade, consists of one 3.7 meter satellite dish transmitting 40 to 50 Mbps bandwidth and supporting 16 time division multiple access (TDMA) nets via six frequency division multiple access links.

B-21. The JNN is deployed at brigade/BCT and division level and is designed to interface with MSE via two simultaneous MSE digital terminal groups supporting voice and data. It connects to a hub (either tactical or strategically mobile hub depending on the network configuration) for further connectivity/access to the DOD's global information grid (GIG) network. The JNN provides circuit switched and internet protocol-based Ku-band commercial satellite capability with up to 7-Mbps in bandwidth.

B-22. The JNN has no transport capability internal to the JNN shelter. It leverages new Ku-band SATCOM equipment along with associated Ku-band TDMA and frequency division multiple access hubs. The JNN also leverages existing capabilities such as Secure Mobile Anti-Jam Reliable Tactical-Terminal (SMART-T), the AN/TSC 85/93 (MSE generation satellite), and high capacity LOS radios for CP to CP links and reach-back links to wide-area services (such as the GIG) and home station. This satellite based transport provides improved mobility and range of individual nodes and CP structures.

B-23. The JNN's tactical local area network (TACLAN) encryptor, KIV-7 and KIV-19 Type 1 encryption can support—

- 48 two wire phone users (SIPR and NIPR).
- 24 internet protocol voice users (SIPR and NIPR).
- 46 internet protocol data users (SIPR and NIPR) (includes 24 data users connected to internet protocol phones).
- Hosts H.323 video conferences and is compatible with the Defense Collaborative Tool Suite.

B-24. The JNN provide top secret/SCI tunneling capability from Trojan Spirit and improves points of presence and mobility in the tactical environment. The JNN is interoperable with U.S. Central Command's combined enterprise regional information exchange network to provide multinational communications support.

B-25. A JNN node is made up of the components shown in figure B-4 and figure B-5, page B-7.

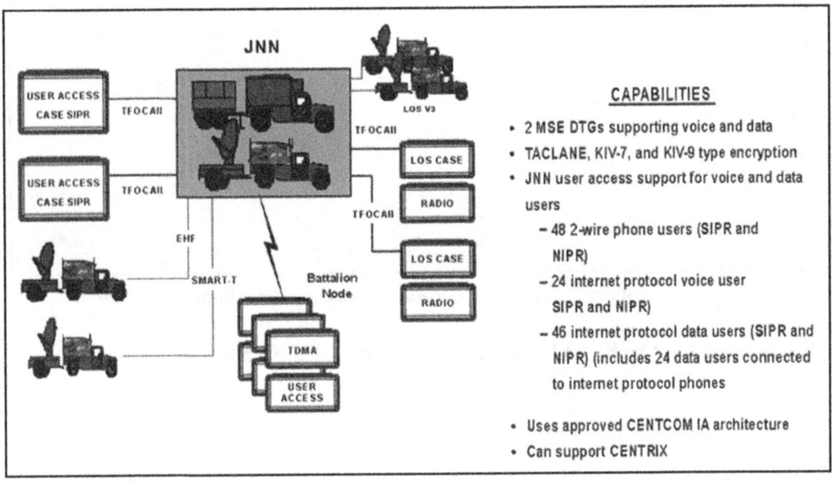

Figure B-4. Joint network node system diagram

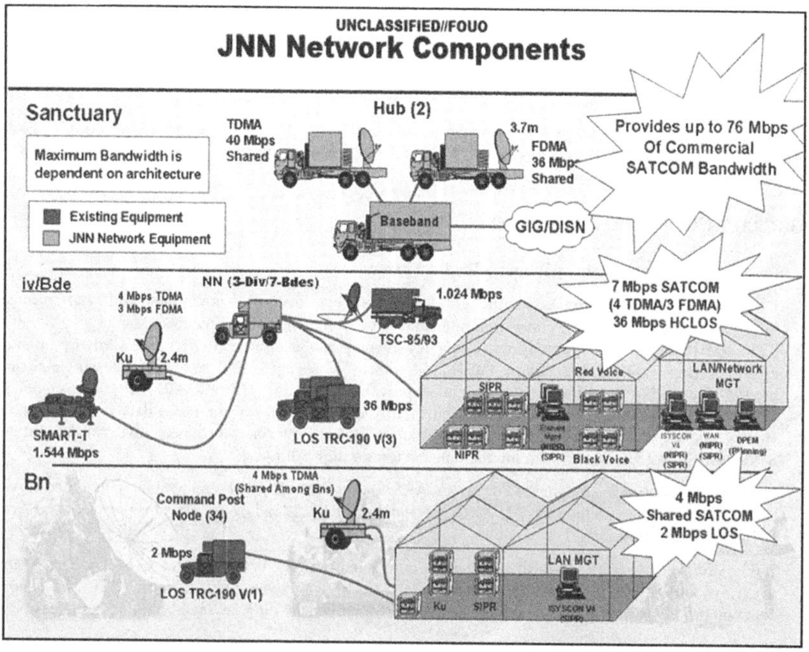

Figure B-5. Joint network node components

B-26. The battalion level or small CP node module is a HMMWV mounted system with a trailer mounted 2.4 meter satellite dish. The module contains a transit cased virtual private network router and TACLAN encryptor security device and provides hub routers for CP and voice over internet protocol (VoIP) phones along with a file server. The battalion level CP module provides 4 Mbps wideband SECRET Internet Protocol Router Network (SIPRNET) data along with VoIP phones to the aviation battalions and links to the JNN through TDMA satellite architecture.

B-27. For unclassified data traffic and interfaces to commercial internet, JNTC utilizes links to DOD's nonsecure internet protocol network (NIPRNET) system. NIPRNET provides for a trusted interface between the DOD intra-net to commercial Internet systems and the World Wide Web through Department of Defense Information Systems Agency designed and maintained demilitarized zone.

B-28. JNTC can provide a link to DOD's SIPRNET. SIPRNET is a worldwide network which allows a secure means to transmit classified data, imagery, and video teleconferencing. SIPRNET can only be accessed by designated secure terminals, and is also available via AKO.

SECTION III – AIRCRAFT AND GROUND COMMUNICATIONS

AIRCRAFT COMMUNICATIONS

B-29. This section discusses capabilities of the following aircraft radios and digital modems:

- SINCGARS (frequency modulated).
- Have Quick II (UHF).

- HF and VHF.
- Transponder; modes 1, 2, 3, and 4.
- EPLRS, BFT, and improved data modem (IDM).
- AN/PRC-112 survival radio and AN/APR-186 (VHF).

B-30. The section also discusses airborne facilitators–such as the UH-60 C2 aircraft, EUH-60 A2C2S, and joint systems—that aid aviation units when relaying communications and challenges to mission communications.

AIRCRAFT COMMUNICATION TYPES

Single-Channel Ground and Airborne Radio System

B-31. SINCGARS is the common battlefield radio system employed by Army ground and aviation forces. It provides secure or plain voice communications over the VHF-FM frequency range of 30.000 to 87.975-MHz at 25-KHz intervals. Its frequency-hopping mode of operation counters enemy jamming efforts. Earlier radio models require the KY-58 to provide secure communications. The single-channel ground and airborne radio system-system improvement program (SINCGARS-SIP) has embedded encryption, an automated GPS interface, and improved data capability for faster data communications. However, even the airborne SINCGARS-SIP requires KY-58 interface for cipher text communications. SINCGARS is a LOS system with limited range at terrain flight altitudes.

B-32. Army aviation's component of SINCGARS is the AN/ARC-201 compatible with other service SINCGARS radios to include the AN/ARC-210 and AN/ARC-222 radios used by other services and Army HH-60L/M air ambulances.

B-33. Aircraft SINCGARS are filled using the automated network control device (ANCD). The AMPS, when available, provides simplified setup of SINCGARS and other radio systems. CTCs have noted common problems with time drift and the need to perform over-the-air rekeying as missions progress.

Have Quick II

B-34. The AN/ARC-164 is a common UHF-AM radio employed by joint aircraft. It provides aviation brigade subordinate units with a means of communicating internally on company battle nets. It also allows interface with sister-service aircraft during JAAT and other joint flight operations. Its frequency hopping mode of operation counters enemy jamming efforts. Like SINCGARS, it is a LOS system with limited range at terrain flight altitudes.

B-35. The AMPS, when available, provides simplified setup of Have Quick II time of day (TOD) and word of day for AH-64D and OH-58D aircraft.

B-36. Units must use Have Quick II in the frequency-hopping mode during training to ensure effective communication during actual operations. Word of the day loading is not difficult, but TOD can be problematic if aircraft lack a Have Quick II/GPS interface. Aircraft without GPS interface can request and accept a GPS TOD from other unit aircraft. In addition, on long operations beyond 4 hours, the TOD begins to drift. A single aircraft, such as the UH-60 C2 aircraft, are then designated as the base point for TOD updates as unit aircraft begin to drop out of the net because of drifting TOD.

High Frequency Radio

B-37. The AN/ARC-220 HF radio system is an NOE, long-range radio system that provides voice and data communication beyond the range of SINCGARS and Have Quick II systems. It operates in the 2 to 29.999-MHz frequency range in 100-Hz steps on 20 preselectable channels for a total of 280,000 possible frequencies. Aircraft not equipped with a 1553 data bus have an additional control display unit for operation of the radio.

B-38. The system has a NLOS range of at least 300 kilometers. The 30 to 100 kilometer range is often the most challenging distance to maintain effective communications.

B-39. Automatic link establishment (ALE) reduces aircrew workload and improves connectivity. In this mode, the caller enters the desired radio address and presses the microphone key. The radio then sounds on the preprogrammed frequency set listening for the best signal. When found, both radios tune to that optimum frequency and a connection occurs. One shortcoming of ALE is third parties do not hear message traffic. If passive listening is necessary and all parties on the net need the same information, the net control station (NCS) chooses the manual or electronic counter-countermeasure frequency-hopping mode. When stations do not rely on each other's reports to perform their mission, ALE is the preferred mode.

B-40. Aircrews can communicate using secure voice or secure data. In data mode, the system can create, edit, and store up to 10 formatted and free text messages of up to 500 characters each. It interfaces with the KY-100 to provide secure communications and the AN/VRC-100 ground radio in aviation ground CPs.

B-41. Secure voice is the primary method of operation for the HF radio in ALE, manual, and frequency-hopping modes. In poor conditions–such as low magnetic flux number, night operations when the ionosphere dissipates, and thunderstorms–aircrews should employ secure data at 300 bits per second. Data transmission increases aircrew workload during flight; the radio stores up to 10 messages in memory, allowing the crew to preload a set of anticipated messages before flight.

B-42. For identical messages with changing location, it often is easier to edit in the new location in an existing memory message than to initiate a whole new entry. In addition, a reduced workload results when commanders use the control display unit's feature permitting HF transmittal of current position with one button press.

B-43. If brigade units have not used HF radios habitually in training, the brigade S-3 should direct HF radio exercises before operations to ensure units use HF to its best advantage.

Very High Frequency Radio

B-44. The AN/ARC-186 is an administrative VHF-AM radio primarily used to communicate with ATS. Normally, it operates in the 116 to 151.975 VHF-AM frequency range. In wired and configured aircraft, it can back up the SINCGARS radio in the same 30 to 89.975 MHz frequency range. It generally lacks a KY-58 interface to provide secure FM communications, and it has no frequency-hopping mode compatible with SINCGARS. The AN/ARC-186 is a LOS radio system with limited range at terrain flight altitudes but greater range at administrative altitudes normally associated with ATS communication.

Transponder

B-45. The transponder enables the helicopter to identify itself automatically when properly challenged by friendly surface and airborne radar equipment. The receiver-transmitter range is limited to LOS transmission. With its frequency of operation in the UHF band range, it is dependent on altitude for range and reception.

Improved Data Modem

B-46. The MD-1295/A is a digital transfer modem that allows equipped aviation forces to exchange complex battlefield information in short, coded bursts. Digital calls for fire are processed through the IDM. The IDM has a preplanned product improvement that incorporates software for processing joint variable message format (JVMF) messages, allowing interoperability with ATCCS and FBCB2.

B-47. A number of joint systems incorporate IDM for data interoperability. The JSTARS CGS, located in brigades and division CPs, also has IDM capability.

AN/PRC-112 Survival Radio

B-48. This small radio, carried in aircrew survival vests, enables downed aircrews to be located by aircraft equipped with the AN/ARS-6 pilot locating system. It receives short, periodic bursts from the AN/ARS-6 and responds with its own coded reply to allow secure location of aircrews. An AM voice mode allows unsecured communication on guard, 282.2 MHz, or two additional UHF channels. The PRC-112A radio has upgraded voice communication security that scrambles voice communication for greater security. Both the PRC-112 and -112A permit voice contact with nearby aircrews if aircraft radios are damaged on impact.

Enhanced Position Location Reporting System

B-49. EPLRS provides a computer controlled communications network which transmits digital information to support TACOPS on the battlefield. EPLRS provides two major functions—data distribution and position location and reporting. It is a secure, jam-resistant, near real-time data communications support system for the five battlefield functional areas of the ATCCS. Because of the real-time unit positioning data supplied by EPLRS, accurate battle management capability increases. This allows the battle commander to not only move forces forward, but to also quickly and accurately counter opposition moves. This information greatly enhances C2 of tactical units by providing commanders with the location of friendly units, a dynamic representation of the FLOT, and abbreviated SITREPs for conditions and identification of adjacent equipped units.

Blue Force Tracker

B-50. BFT, a component of the FBCB2 system, is an efficient tool commander use for SA, airspace deconfliction, and C2. As a C2 tool, BFT allows the commander to track aircraft locations and provides an alternative means of over-the-horizon communications to meet this challenge. It also fills communication gaps by providing the capability to pass text messages between stations. Code words and similar short text transmissions are easily passed to supplement, or even replace, radio calls.

B-51. In planning, BFT enhances C2 by enabling the COP to readily be shared between headquarters and between aircraft. Graphic control measures such as PZs, flight routes, restricted operations areas, LZs, and FSCMs can be developed, plotted, and shared with other BFT-equipped units as a computer-graphics overlay file. These graphics can be downloaded at each BFT station (whether stationary, aircraft-, or vehicle-mounted) to enable viewing by crews.

Utility Helicopter Command and Control Console

B-52. UH-60 aircraft equipped with the AN/ASC-15B C2 console provide users with in-flight SA and communications access. The modified console provides SINCGARS, Have Quick II, HF, VHF-AM, and SATCOM. Console systems are supported by aircraft power and internal aircraft antennas. The aircraft has one SINCGARS 201 radio and three AN/ARC 210 multimode radios capable of operation on SINCGARS FM, UHF, or VHF frequencies. This permits the capability to simultaneously operate the command network and monitor the O&I or higher headquarters command networks. It provides operators with a means of choosing between either active SINCGARS communication or retransmission. Retransmission of Have Quick II and VHF-AM is also possible with the system.

B-53. The aircraft console contains radio sets, console controls, and six internal communication system boxes. In the rear, four additional internal communication system boxes and a map board allow up to 10 personnel to monitor the console's radio systems. The C2 console's lights are compatible with NVG. It is the supported unit's responsibility to provide a trained console operator; the crew chief is not trained to perform this function.

B-54. The C2 console can operate in the ground mode. In this configuration, the console can remain mounted on the aircraft or be dismounted. In the ground mode, the C2 console requires generator power and external antennas. Four trained personnel can remove the console from the aircraft in 1 hour. Figure B-6, page B-11, shows the aircraft configuration.

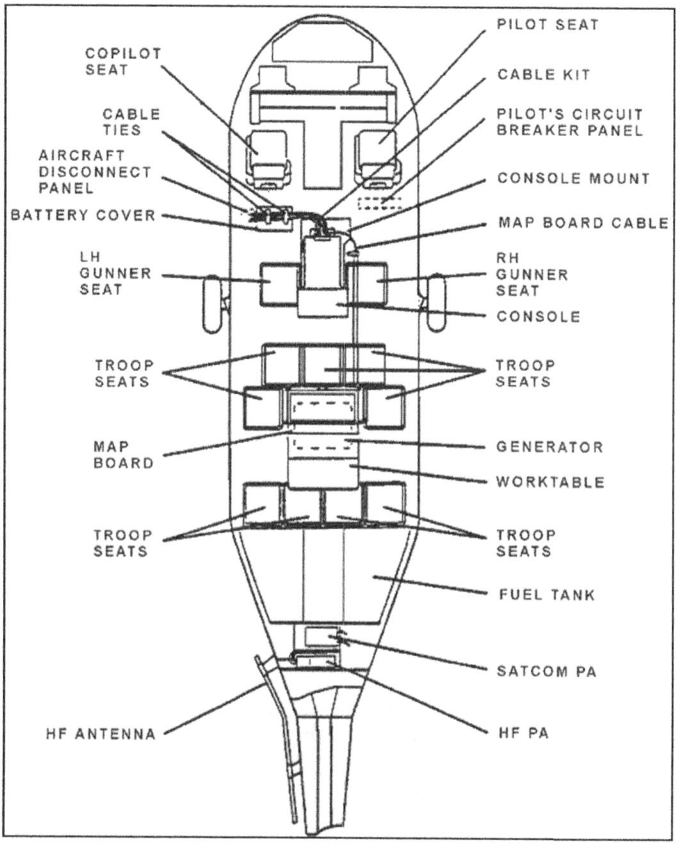

Figure B-6. UH-60 command and control aircraft configuration

Army Airborne Command and Control System

B-55. The A2C2S is a UH-60 based C2 system (referred to as the EUH-60) that serves as an airborne TAC CP. Through its onboard MCS, ASAS, advanced field artillery tactical data system (AFATDS), air and missile defense work station (AMDWS), combat service support control system (CSSCS), and FBCB2, A2C2S provides continuous battlefield SA. It also is the source of digital information for nondigitized aircraft supporting the operation. A2C2S provides maneuver commanders—from ARB/ARS to EAB—with on-the-move C2. The system supports three major operational functions—mission planning, mission execution, and mission support. Its primary function is to monitor execution of current operations while the main CP focuses primarily on planning future operations.

Capabilities

B-56. A2C2S enables the commander and his staff to traverse the OE to critical places at critical times. The commander and staff can perform all battle command and coordination functions from A2C2S. It has simultaneous multiband voice and data channels and dynamic visual battlefield SA and C2 via command, control, communications, computers, and intelligence connectivity. A2C2S provides access to the TI to manipulate, store, manage, and analyze SA information, intelligence data, mission plans, and mission progress data to support the C2 decisionmaking process. The system has triservice interoperability and is compatible with NATO, civil aviation, maritime, and law-enforcement communications (figure B-7).

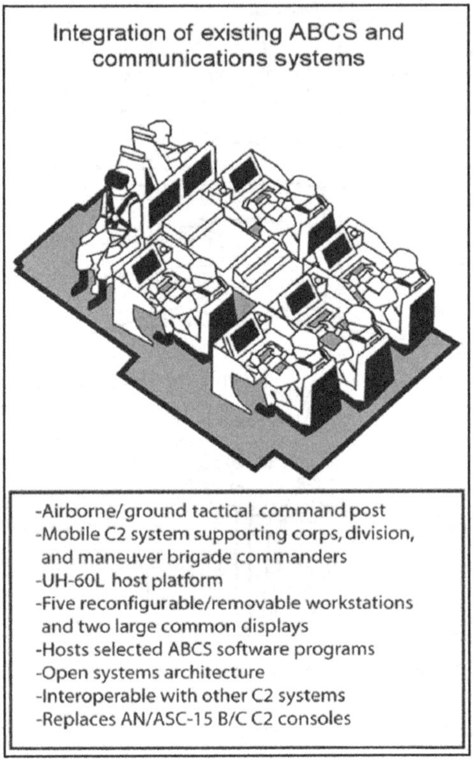

Figure B-7. Army airborne command and control system configuration

Features and Performance

B-57. A2C2S provides—

- Robust LOS and NLOS communications through SINCGARS advanced system improvement program, SATCOM demand assigned multiple access, Have Quick II, EPLRS (friendly positions), NTDR (SA), and HF.

- GPS for present position and standard National Geospatial-Intelligence Agency (NGA) maps with overlays for a complete picture of the battlefield.
- Automated display of COP information and C2.
- Five automated, reconfigurable, and removable workstations, a command database, and two large common displays; each workstation incorporates a keyboard, monitor, and audio communications unit.
- Real-time OE control and monitoring.
- Common displays.
- Enhanced control of battle.
- Digital connectivity with all ABCSs.
- Standard communications and information security.
- Airborne and ground operational modes.

Interfaces

B-58. A2C2S interfaces with—

- JSTARS and SATCOM.
- Maneuver CPs.
- CH-47Fs, AH-64Ds, and OH-58Ds.
- M1 main battle tanks and M2/M3 cavalry fighting vehicles.
- MLRS.

Operation as a Ground Command Post

B-59. When operating as a ground CP, the preferred power source is commercial power. If commercial power is not available, a generator is the next preferred power source. If external power is not available, aircraft power is required. Extended ground times may require a ground power unit, which may be brought in via sling load or tactical ground vehicle, such as a HMMWV with a generator kit.

Employment

B-60. The IM capabilities of A2C2S are focused on controlling operation execution; planning capability is limited. Mission data are transferred to A2C2S from the digital CP to bring it to the start of a mission operational status.

Information Flow

B-61. The ATCCSs are primarily top-down planning tools. Once the execution phase begins, the primary flow of information is bottom-up via FBCB2. A2C2S draws real-time data from broadcast sources to determine changes to the enemy situation during the mission's execution phase. The intelligence information provided by ASAS is an analyzed and formal product. Intelligence information A2C2S receives from tactical related applications, tactical data information exchange-broadcast, and tactical information broadcast service broadcast sources is raw data (figure B-8, page B-14).

Battlefield Employment

B-62. A2C2S expands the battlefield by providing the means to exercise C2 and gather tactical information in support of a mission while on the move. From A2C2S, the commander and staff influence the battle via direct exchange of voice and digital information with units conducting the mission. They simultaneously develop the situation beyond the range of their unit's sensors and shooters by accessing broadcast intelligence sources.

Covering Force and Unassigned Areas

B-63. A2C2S enhances lethality during covering-force missions and shaping operations in unassigned areas by moving its command forward to maintain contact with the maneuver forces. From A2C2S, the commander and staff can synchronize deliberate and hasty artillery fires. A2C2S has a direct link to artillery, including the Army tactical missile system. However, direct linkage is not necessary for direct FS or priority of fires.

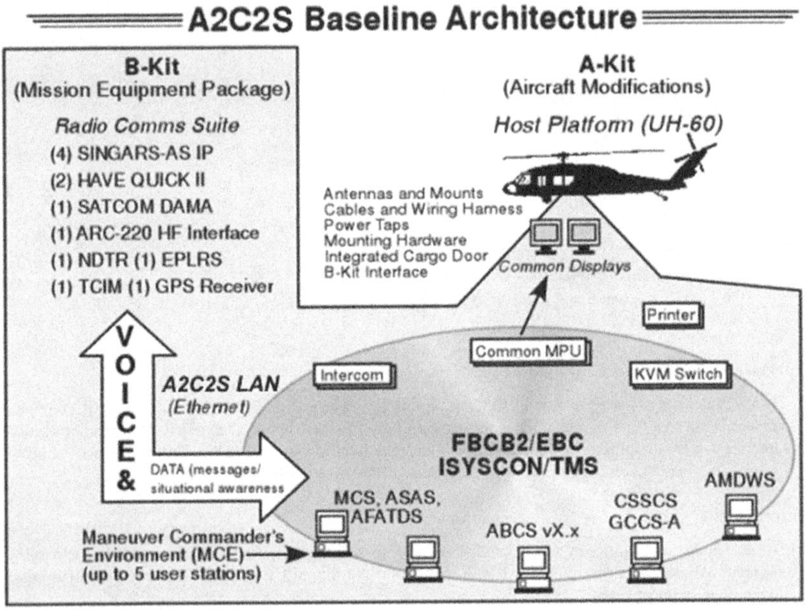

Figure B-8. Army airborne command and control system information flow

Assigned Areas

B-64. Integral activities during operations in assigned areas include maneuver, close combat (including FAS), indirect FS, forces and sustainment of committed forces and command, control, communications, and intelligence. Aviation organizations may be employed as a security or reserve force in the security or main battle area. A2C2S gives the commander a clear picture of the close battle and allows him or her to coordinate and synchronize maneuver and fires. Linked with other automated systems, A2C2S can pull information on demand allowing the commander to operate at his own tempo without the information delays characteristic of traditional reporting methods.

Sustainment Areas

B-65. The aviation brigade gives the division commander a highly mobile and lethal combat force to counter a Level III incursion into the sustainment area. As a maneuver headquarters, the brigade may be tasked as a TCF to respond to a significant threat. A2C2S provides a flexible and highly mobile tactical CP to control operations.

Stability Operations

B-66. During stability operations, the system provides connectivity to special operations, C2, embassy, law enforcement, maritime, civil, and/or other humanitarian information and communication networks. A2C2S can also improve the ability of local, state, and federal agencies to communicate and coordinate in a crisis environment such as a hurricane or forest fire during civil support operations.

AIRBORNE RELAY

B-67. Some operations in unassigned areas have priority to justify communications relay as a means of overcoming difficulty in communicating. If allocated, the C-12 may perform HF relay or even SINCGARS and Have Quick II relay if the threat permits flight within range of those radio systems. The AWACS, E-8 JSTARS, C-130 airborne battlefield C2 center, EA-6, airborne FACs, participating deep JAAT and AI, or other joint aircraft may be available to relay HF, Have Quick II, and in some cases, SINCGARS communications. EPLRS capabilities on the A2C2S aircraft allow automated relay of data communications. In addition, future UAS may have retransmit mission capabilities for FM command nets. Table B-1 illustrates the potential for relay with higher-flying aircraft if coordinated by staff members in advance.

Table B-1. Joint aircraft potentially interoperable for communications or relay

Comms/Relay Capable	*C-12*	*E-3 AWACS*	*E-8C JSTARS*	*C-130 ABCC C*	*EA-6B*	*FAC*	*AI/JAAT*
SINCGARS		X	X	X	X	X	
Have Quick II	X	X	X	X	X	X	X
HF	X	X	X	X	X	X	X
EPLRS		X	X	X			X F16 Block 30
IDM			X		X		X F16C F16D

AIRCRAFT COMMUNICATIONS CHALLENGES

B-68. The primary challenge to aircraft communication is the combined effect of terrain-flight altitudes and operational distances between aircraft and their CPs. The HF radio is the primary materiel solution to the NOE communications requirement and need to communicate over greater distances. In addition, unlike SINCGARS, only a single HF radio is available on most aircraft. These constraints relegate the HF role to a secondary communications system available when other communications are impossible.

B-69. Army aircraft share common radio systems and have communications interoperability. One exception is the OH-58D that lacks HF capability. The AH-64A and CH-47D also have a single SINCGARS radio. This situation prohibits commanders/staffs from simultaneously monitoring both command and O&I nets. It also inhibits routine data communication. Table B-2 compares Army aircraft communications capabilities.

Table B-2. Aircraft communications interoperability

Tactical Aircraft Communications	AH-64D	AH-64A	OH-58D	UH-60AL	CH-47D	HH-60/M
AN/ARC-186 VHF-AM/FM	X	X*	X	X	X	
AN/ARC-201 VHF-FM (SINCGARS)	X (2)	X (1)	X (2)	X (2)	X (1)	X (2)

Tactical Aircraft Communications	AH-64D	AH-64A	OH-58D	UH-60AL	CH-47D	HH-60/M
AN/ARC-220 (HF)				X	X	X
AN/ARC-164 (Have Quick)	X	X	X	X	X	X
AN/ARC-222 VHF-AM/FM						X
MD-1295/A (IDM)	X		X			X
*Same antenna for VHF-AM and FM 2 commo.						

GROUND COMMUNICATIONS

ELECTRONIC

B-70. For unclassified traffic, units can utilize several commercial electronic communications resources. These resources include—

- Email (both civilian and military addresses).
- Instant messaging.
- Text messaging.

B-71. SIPRNET is a worldwide router-based network which allows a secure means to transmit classified data, imagery, and video teleconferencing. SIPRNET can only be accessed by designated secure terminals, and is available on AKO.

WIRE

B-72. Wire communication should be the primary means of communicating within the CP areas when practicable. Subordinate and attached battalion main CPs run wire to the aviation brigade main CP. Wire across roads either overhead or through culverts and bury as soon as possible to hinder enemy tapping.

GROUND SINGLE-CHANNEL GROUND AND AIRBORNE RADIO SYSTEM

B-73. Ground SINCGARS is the primary C2 network within the brigade and corps/division. It is also used for O&I and A&L networks. Some systems require KY-57 for security. The newer SINCGARS-SIP has data rate adapters and encryption embedded. On vehicle-mounted SINCGARS, the user looks for /A after the SINCGARS numerical designation to identify systems with integrated COMSEC. The ANCD or AMPS allows loading of SINCGARS and IFF information.

GROUND HAVE QUICK II

B-74. This ground radio allows communications with Have Quick II UHF-AM airborne radio systems. It includes a portable GPS for aligning TOD and a KY-57 for secure communications. It is backward compatible with first-generation Have Quick systems and non-Have Quick UHF-AM radios. It is compatible with , Navy, and Marine Corps Have Quick II systems, but LOS constraints may hinder communication with joint systems from the ground.

MOBILE SUBSCRIBER EQUIPMENT

B-75. The MSE network architecture forms a nodal grid system capable of providing multiple communication paths between node centers throughout the Army's AO. Multiple paths via LOS radios and tactical satellite links between node centers ensure a high degree of system survivability. Small extension nodes (SENs) are connected to node centers providing communications support to smaller units. MSE employs ground LOS, troposphere scatter, and satellite transport. The SMART-T, if available, can also be used to provide satellite range extension for the MSE network. MSE provides the aviation brigade the ability to maintain connectivity with dispersed aviation units engaged in operational missions.

B-76. MSE is designed to provide a connection between the aviation brigade's main CP and higher headquarters as well as providing support to the brigade's organic units as assets allow. SENs and radio access unit support provide both MSE telephone and mobile subscriber radio-telephone coverage for the aviation brigade and battalion CPs.

B-77. For unclassified data traffic and interfaces to the commercial internet, MSE utilizes links to the DOD's GIG which carries the NIPRNET. NIPRNET provides a trusted interface between the DOD intranet to commercial internet systems and the World Wide Web through Department of Defense Information Systems Agency designed and maintained gateways called demilitarized zones. These gateways ensure the DOD network maintains its integrity and guards against computer attack.

B-78. MSE also provides a link to the DOD's SIPRNET carried on the GIG. SIPRNET is a worldwide network which allows a secure means to transmit classified data, imagery, and video teleconferencing. SIPRNET can only be accessed by designated secure terminals, and is also available on AKO. SIPRNET is a closed network; however actions have been taken to interface SIPRNET with the Department of Homeland Security's Homeland Security Network. Refer to FM 11-55 for a more complete description of MSE equipment, architecture, and operations.

NODE CENTERS

B-79. Node centers serve as access points for large nodes, SENs, and remote access units and are linked together to form the backbone of the MSE network. For a typical division, the grid is made up of 4 node centers. A typical corps has 42 node centers. Node centers can be emplaced up to 40 kilometers apart.

B-80. SENs provide communication network access to smaller units such as battalion and brigade CPs. Access for static units is by wire.

GROUND HIGH FREQUENCY

B-81. The AN/VRC-100, coupled to the KY-100, provides secure communications with airborne HF radios. The VRC-100 and aircraft ARC-220 have virtually identical components packaged differently.

B-82. Because HF radio waves bounce off the ionosphere, short-range HF is difficult to direction find and jam. If jamming does occur in the ALE mode, ALE simply finds a better frequency. If jamming occurs in manual mode, the NCS may not be able to announce a mode switch to all stations. Aircrews that lose HF communications must exhaust other possibilities before assuming jamming is the problem and switching to the electronic counter-countermeasures frequency-hopping mode without net notification.

B-83. Antenna selection and angle are critical to effective communication using the high frequency radio. Table B-3 illustrates different antenna configurations and applications. Only the FANLITE near-vertical incident skywave antenna comes standard with the radio system.

Table B-3. Antenna configuration effect on operational range

Antenna Type	*Radiation Pattern*	*Antenna Takeoff Angle*	*Value to Operations*
32 ft whip, vertical	Omnidirectional	45 degrees with ground radials installed	Fair at medium range
16 ft whip, vertical	Omnidirectional	Vertical to 45 degrees	Poor, for mobile use only
Standard FANLITE sloping or horizontal	Near vertical	45 degrees to horizontal	Good at short range
Resonant di-pole, horizontal	Bidirectional	45 degrees to horizontal	Good at medium range
Log periodic	Unidirectional	Where pointed	Very good at long range when pointed on the horizon; very good at short range when vertical
Yagi	Unidirectional	Where pointed	Good at long range when pointed on the horizon; good at short range when vertical

B-84. Besides antenna considerations, frequency selection is another critical variable for effective HF communications. HF radio frequencies for effective short-range (30 to 100 kilometers) communications are usually below 8 MHz. The FANLITE antenna works better and the ground wave is longer at lower HF frequencies. However, the corps or division signal office typically assigns frequencies without considering these parameters. The brigade S-6 must ensure the higher headquarters signal office is aware of optimal aviation HF frequencies.

B-85. At night, the ionosphere dissipates, resulting in less reflection of HF radio waves. When this situation occurs, relay over a longer path may prove effective. A more distant station may receive the HF signal better than a close one. Ground HF operators should have a list of frequencies and call signs to contact other distant aviation brigades or stations that can relay C2 information.

B-86. In the ALE mode, if the radio channel is inactive for a period of time, the radio reverts to the scan mode and another ALE sequence must occur to reconnect. To prevent this situation, stations operating in the ALE mode sound periodically to retain a good frequency for communication. This sounding ensures an ALE connection is already in place, thereby saving time when a message must be sent. Radios can be set up to automatically sound at a periodic rate. The ground HF radio operator generally can perform sounding to reduce aircrew workload.

Air Traffic Services Communications

B-87. Air traffic control radios are available for AC2, limited flight following, and localized control of inbound and outbound aircraft. Radios also permit recovery of aircraft that experience inadvertent IMC. These systems may provide brigade commanders with a backup means of communicating with units, although this should not be their primary mission. Commanders must recognize radios emit unique signatures and locating them to the brigade CP must be balanced with the knowledge some enemies can identify and target signature location. Another option available to brigade commanders is employment of better ATS antennas used with other tactical radios.

B-88. The TAIS provides fully automated capability to support airspace management at theater, corps, and division level. TAIS is fully integrated with ABCS. When used with other ABCS, TAIS provides automated AC2 planning and airspace deconfliction. The tactical terminal control system, AN/TSQ-198, provides tactical ATS capabilities in more austere environments. It can also provide backup communications capabilities at aviation CPs or in unassigned or sustainment areas.

Ground Satellite Communications

B-89. Different SATCOM ground systems may be available to aviation brigades. For effective use, CP locations must permit LOS between the dish antenna and geosynchronous satellites. For instance, a CP location next to a mountain or among tall trees may obstruct SATCOM LOS. To prevent SATCOM bleed-over, a minimum of 10 MHz frequency separation should exist between outgoing and incoming signals.

B-90. Common SATCOM systems include the PSC-5 Spitfire and AN/PRC-117F. These systems include SINCGARS and Have Quick capability. The SMART-T is a larger SATCOM system that interfaces with military strategic and tactical relay satellites for data transfer at low and medium rates to extend the MSE network range.

B-91. Units should avoid over reliance on SATCOM for longer-range communications during large-scale conflict as channels can become oversubscribed. In addition, SATCOM may not be a viable solution in certain latitudes and areas of the world where geosynchronous satellite coverage is sparse.

Portable Hand-Held Two-Way Radios

B-92. The walkie-talkie radio is nonsecure and operates in the 138 to 160 MHz FM range. The Army version (AN/PRC-127) has fourteen available channels, and the frequency is set from an integral keypad. These radios provide personnel with a low-power means of localized communication.

Commercial Telephone Lines and Cellular Telephones

B-93. In many areas, commercial telephone lines and cellular phones can support nonsecure voice and data communications or prompting between parties to attempt communications using secure means.

Video Teleconference

B-94. Tactical VTC capability resides in several brigade CPs, but not yet at battalion level. A VTC provides the capability to communicate visually with audio between several linked VTC stations.

Visual and Sound Communications

B-95. Visual card systems, landing lights, hand-and-arm signals, flags, pyrotechnics, and other visual cues can provide simplified communications when radio transmission may not be possible or tactically sound. Visual cues are especially valuable in FARP, sling-load, and ATS operations near AAs. Audio cues are another possibility, such as for alert of chemical attacks, but around operating vehicles and aircraft audio signals may prove inaudible.

Messenger

B-96. Ground and air messengers may transport hard-copy messages and larger documents as part of a regularly scheduled shuttle between CPs, field trains, and higher and lower headquarters. An alternative to dedicated messengers is delivery with ground and air of supplies such as meals delivered to a tactical CP. Messengers may deliver combat plans and orders, written coordination and control measures, graphics, logistics requests and estimates, or other extensive documents that would consume excess time to send electronically.

SECTION IV – AIRCRAFT COMMUNICATIONS EMPLOYMENT

B-97. SINCGARS is the primary combat net radio. Airborne commanders operate on the command net. Reports are sent on the O&I network. Logisticians and FARPs operate on the A&L net. Have Quick II supports internal communication between aircraft at the company level and provides a means of communicating with any joint air systems participating in the mission. HF communications enhance terrain flight communications with distant CPs. If UH-60 C2 system-equipped or EUH-60 A2C2S aircraft are available, SATCOM provides another long-distance communication option. Units minimize voice communications by employing brevity codes and digital data communications.

ATTACK RECONNAISSANCE BATTALION/SQUADRON

B-98. Longbow-equipped units have secure FM1 and FM2 SINCGARS capability to simultaneously operate on two nets. One radio can habitually operate for voice and the other for data.

B-99. Have Quick II voice mode or IDM data transfer facilitates company and platoon internal communication. Designated aircrews can make reports to battalion on the O&I SINCGARS net, while keeping the company commander aware on an internal Have Quick II O&I net. The HF radio is available as a secondary means of voice or data communication with the battalion. AH-64A units have neither dual secure FM radios nor an IDM capability. These units can employ HF secure data communication as an alternative to FM2 secure/IDM.

B-100. OH-58D aircraft have secure FM1 and FM2 SINCGARS, Have Quick II, and VHF capability. The following is a preferred means of internal battalion communications:

- FM1 (secure) battalion command net (battalion commander, XO, S-3, and company commanders).
- FM1 (secure) platoon command net.
- FM2 (secure) digitized O&I network/supported unit/FS net.
- Have Quick II (secure) company command net.
- VHF (nonsecure) coordination net for all elements.

B-101. The FM2 may be designated as a digital SA network for IDM-transmitted SPOTREPs, SITREPs/STATREPs, and BDA reports. These digitized reports are sent via FM2 directly to battalion and company FBCB2-equipped vehicles.

ASSAULT HELICOPTER BATTALION

B-102. Battalion UH-60 aircraft missions range from single ship air movement to major air assaults involving multiple aircraft. As with other units, the primary combat net radio is SINCGARS, which is employed for command, O&I, and A&L nets. For intra-aircraft communication, units use Have Quick II. In the absence of a SINCGARS/IDM capability and given typical air assault distances, HF is a secondary and often crucial communications tool for maintaining contact with distant CPs. To minimize voice traffic on air assaults, AMCs employ HF ALE data transmission with preloaded short messages for anticipated reports to the rear. These could include—

- Staging phase: arrival PPs, crossing phase line, arrival PZ, executing bump plan, PZ unsecured, executing/arrival alternate PZ, request maintenance, enemy contact, and downed aircraft.
- Air movement phase: arrival start point/RP, reporting airspace control plan 1, executing bump plan, executing/arrival alternate LZ, request maintenance, unanticipated enemy contact, downed aircraft, and request for aeromedical evacuation.

B-103. Single ship air movements can occur at extended distances. Unit CPs can communicate changes in pickup and drop-off points and other en route changes using the HF ALE mode to assure communications contact.

GENERAL SUPPORT AVIATION BATTALION

B-104. The GSAB has EUH-60 A2C2S aircraft. Ground brigade commanders and staffs employ A2C2S, as required, without interference from aircrews. Aircrews may be asked to monitor certain SINCGARS nets on aircraft radios and relay key messages to staff members in the sustainment area. This requirement and distance involved may require aircrews to use HF communication to maintain contact with the command aviation battalion CP or relay messages for supported commanders if C2 system HF radios are in use or ineffective.

B-105. A secondary mission of EUH-60 A2C2S aircraft is C2 of some aviation brigade missions such as operations in unassigned areas and air assaults. In these missions, the aviation brigade commander and selected staff may employ the A2C2S aircraft as a tactical CP. Relative proximity to mission aircraft facilitates SINCGARS voice and IDM data transmission between brigade and battalion commanders. The availability of HF and SATCOM ensures long-distance communications with the division or corps CP.

B-106. Heavy helicopter missions are frequently single ship long-distance operations and require HF for communications with the battalion CP. Some units employ multiple CH-47s for air assaults to move artillery, HMMWVs, and other key mission equipment. These missions require organic SINCGARS capability to communicate on assault battalion nets; however, only one SINCGARS is generally available. Have Quick II provides internal communication between CH-47s.

AVIATION BATTALION TASK FORCE

B-107. An ABTF forms and deploys for missions not requiring an entire aviation brigade but supporting a broad spectrum of aviation missions. The AH-64D, OH-58D, and HH-60L/M have IDM capability for data communications; the AH-64A, UH-60A/L, and CH-47 aircraft do not. All aircraft share SINCGARS, HF, and Have Quick II interoperability with the exception of the OH-58D, which lacks HF capability.

B-108. For some missions requiring extensive digital communications, such as attack, only IDM-capable OH-58D and AH-64D aircraft may participate. However, OH-58D aircraft may be task-organized with non-IDM AH-64As. During reconnaissance and air assaults, all aircraft may participate. TF commanders require cross-trained staff personnel and possibly A2C2S aircraft to C2 the TF.

SECTION V – DIGITIZATION

B-109. Force projection, split-base operations, information warfare, and joint or combined operations are doctrinal concepts for warfighting. Crucial to these capabilities is the effective information flow to support warfighting throughout all phases of an operation (figure B-9, page B-21).

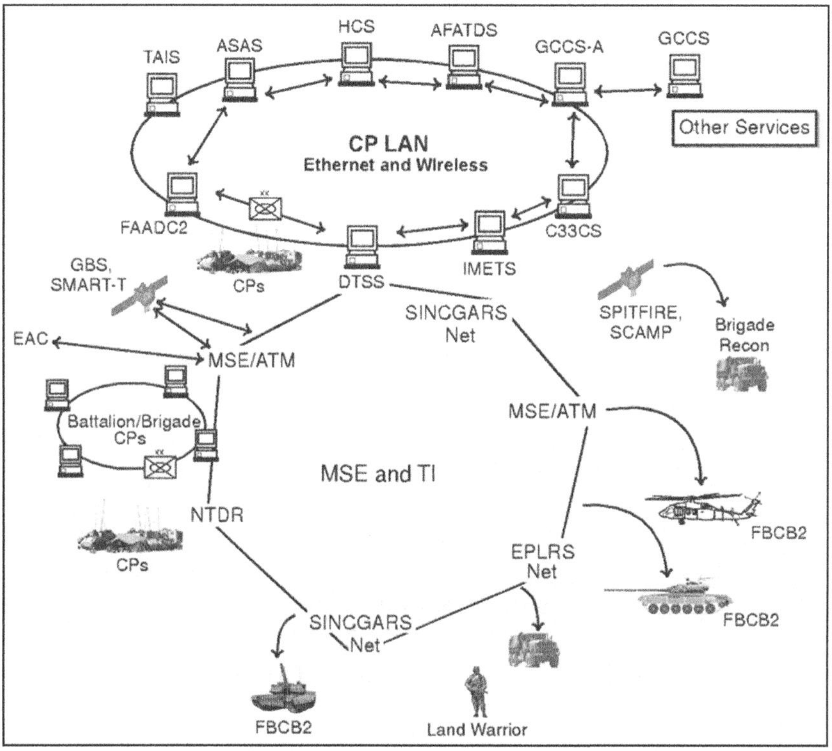

Figure B-9. Digitized communications

ARMY BATTLE COMMAND SYSTEM

B-110. ABCS provides rapid and reliable information nets enabling the Army to project and protect the force, gain information superiority, determine the OE, conduct decisive operations, and sustain the force. It provides real-time and near-real-time information that enables sound decisionmaking inside the enemy's decision cycle.

B-111. ABCS is a collection of IM systems that assist the commander in exercising C2 and gaining SU. ABCS permits him or her to apply judgment more productively, use command presence more efficiently, to develop and disseminate his vision effectively, and understand better the dynamics of war (in general) and specific operation (in particular).

B-112. ABCS provides a visual means to see friendly and enemy forces and the ability to arrange and maneuver forces to accomplish missions. The ABCS components assist in answering the following questions:

- Where am I?
- What is my status?
- Where are the other friendly units?
- What is their status?
- Where is the enemy?
- What is the enemy's status?

B-113. ABCS Version 6.4 (figure B-10) provides for key technology enhancements of the current ABCS to include integration and dissemination of terrestrial and satellite based FBCB2 BFT and C2 data, transitions battle command systems from specialized workstations to commercial-off-the-shelf laptops, and introduces net-centric, XML-based publish and subscribe architecture.

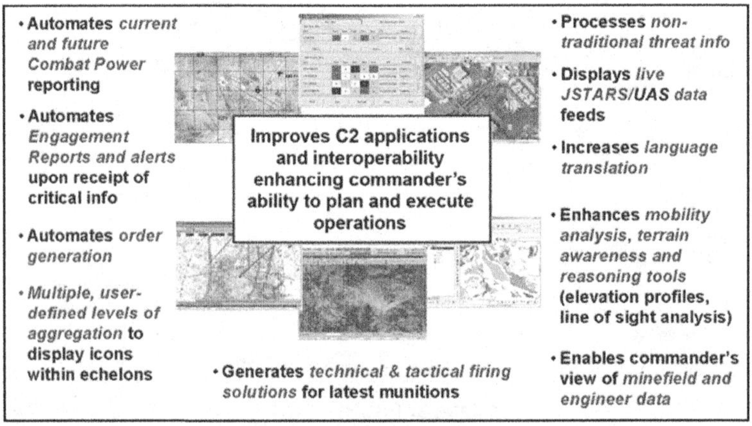

Figure B-10. Army battle command system 6.4 operational enhancements

B-114. Digitization capability is an evolutionary process occurring over many years. When capability is incomplete, the challenge is to devise ways to mix traditional, manual methods with the automated systems that permit more rapid planning and synchronized execution. As always, these guidelines should be applied to a degree that complements the existing level of automation.

Common Pictures

B-115. The terms COP and common tactical picture (CTP) are often used interchangeably, but have distinctly different definitions.

- The COP is an operational picture tailored to the user's requirements, based on common data and information shared by more than one command; the COP facilitates collaborative planning and assists all echelons in achieving SU, which helps to synchronize execution.
- The CTP is an application available on ABCS computers and supporting systems; it uses a common mapping background, is accessed through a common user interface, and displays information shared from the joint common database (JCDB). The CTP is dynamically updated as data changes in the JCDB.

B-116. Examples of COP overlays are force disposition, enhanced by overlaying the operational overlay; FS overlays; and AC2 overlay. Additional information is available at the description of each system.

B-117. ABCS assists by providing a COP of the OE through timely presentation of information in various types of formats including voice, data, imagery, graphics, and video. The operational picture also provides—

- Access to planning documents.
- STATREPs.
- Timely, automatic warnings of air, missile, and CBRN attacks.

B-118. Although each battlefield automated system (BAS) of ABCS makes contributions that support its own Army WFF-oriented tasks, the key contribution of ABCS is as an interoperable system of systems. The synergistic capabilities of ABCS allow commanders to reach across the Army WFFs to request, select, and evaluate data from diverse resources to create relevant information. The COP begins with a common map background against which a commander can display a variety of information such as—

- Friendly locations and graphic-control measures.
- Enemy units and equipment.
- FSCMs, range fans, and targets.
- Air tracks and tactical ballistic missile tracks.
- Logistics status and joint information.

B-119. The COP includes all relevant elements such as—

- Army units.
- Joint, allied or coalition forces.
- Enemy forces.
- Neutral elements.
- Unknown forces.

B-120. Each user can tailor his COP to show as little or as much information as he or she requires. ABCS's essential contribution to C2 is it provides identical, shared data. ABCS enhances warfighting in the following ways:

- Accelerates the MDMP, preparation of estimates, COA development, wargaming, and orders production and dissemination.
- Assists in gathering and displaying relevant information while filtering unnecessary data.
- Allows dissemination of information in near-real time and minimizes latency of information exchanges.
- Facilitates synchronization of sustainment by increasing opportunities for real-time coordination.
- Exploits digital map data and terrain-analysis products.
- Facilitates rehearsal and training through compatibility with current and future simulation and simulation systems.
- Enhances interoperability through commonality of task procedures.
- Provides data access to the commander in austere environments through reach-back capability.

COMMON SERVICES

B-121. ABCS provides collaborative tools, training programs, and applications.

Collaboration Tools

B-122. Collaboration tools include—

- VTC, whiteboard, and shared applications.
- Messaging.

- File transfers.
- Calendar creation/scheduling.
- Task management.
- Internet browser.
- Database query tools.

Training Programs

B-123. These provide training and simulation capabilities for individual and collective training events.

Applications

B-124. Common applications include word processor, spreadsheet, and presentation/graphics programs. Document interchange services support document exchanges between heterogeneous computer systems using common file formats. The operational picture application creates a shared picture of the OE. The planning application automates aspects of the MDMP and enables parallel and collaborative planning.

Army Battle Command System Communications Net

B-125. Connectivity is provided by tactical communications systems—MSE, NTDR, SINCGARS, and EPLRS. The ABCSs within brigade, division, and corps CPs are supported by a wide area network (WAN) and LAN switch/router architecture (figure B-11, page B-25).

Other Digital Systems

B-126. Additional systems interfacing with ABCS may include A2C2S, digital topographic support system/quick response multicolor printer, and CGS.

SUBSYSTEMS TO THE ARMY BATTLE COMMAND SYSTEM

B-127. ABCS consists of information technology applications, nets, and communications enabling data exchange subsystems throughout the force. Each subsystem supports and provides information to other systems to improve battlefield SU. By integrating the ABCS components to a JCDB, the COP can be viewed at any workstation according to the commander's specific requirements. In addition, ABCS subsystems provide an array of specialized capabilities and applications for units at all levels.

B-128. ABCS consists of the following subsystems:

- GCCS-A.
- FBCB2.
- TAIS.
- Digital Topographical Support System (DTSS).
- Integrated meteorological system (IMETS).
- ATCCS.

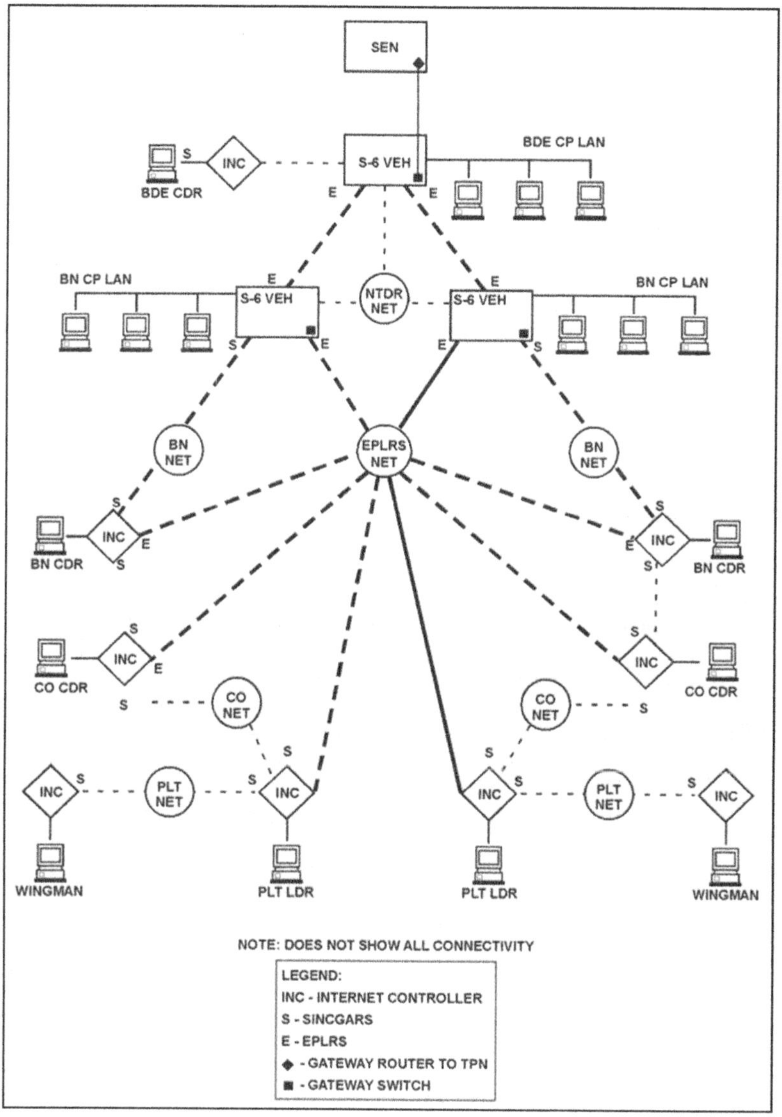

Figure B-11. Example of an Army battle command system communications net

Army Battle Command System and the Common Tactical Picture

B-129. Figure B-12 shows the ABCS's input that forms the CTP.

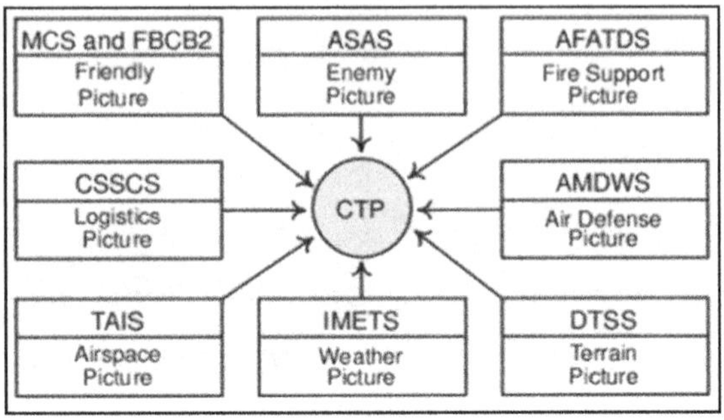

Figure B-12. Common Tactical Picture

Global Command and Control System–Army

B-130. GCCS-A is the Army hardware and software directly supporting Army implementation of the joint global command and control system (GCCS). It supports monitoring, planning, and execution of joint, combined, and Army operations for theater GCCS-A ensures Army access to key information within the joint realm such as force tracking, host-nation and civil affairs support, theater AD, targeting, psychological operations, C2, logistics, and medical and personnel status. In turn, this information supports corps-level planning, execution, and monitoring of mobilization, deployment, sustainment, and redeployment.

B-131. A GCCS-A system is at the corps main and tactical CPs.

Capabilities

B-132. The commander's force analyzer provides current TPFDD. This information is key to planning movement of forces and monitoring unit status and availability.

B-133. The logistics analyzer allows planners to forecast resources needed in various combat situations.

B-134. GCCS-A shares the client-server architecture common operating environment with the joint GCCS for general functions of teleconferencing, messaging, file transfers, office automation, utilities, and system administration.

Force XXI Battle Command Brigade and Below

B-135. FBCB2 provides C2 and SU to the lowest tactical echelons. It supports OPCON chiefly through the transmission and receipt of orders, reports, and data via combat messages. FBCB2 employs position navigation and reporting capability to depict and transmit the unit's own location. FBCB2 can also access other friendly units' locations, as well as intelligence, to show the friendly and enemy picture in near-real time, even while on the move.

B-136. FBCB2 is found on platforms from the commander to the Soldier level.

Capabilities

B-137. FBCB2 assists SU by telling the user his location and locations of other friendly forces, observed enemy forces, and reported battlefield obstacles. The user can adjust his picture of the battlefield by selecting which overlays, graphics, and icons are shown. Unit displays can be altered by grouping icons according to unit type or echelon.

B-138. FBCB2 automates frequently used urgent messages for reporting the enemy, requesting MEDEVAC, CBRN attack, call-for-fire, cease fire, and unit situation reporting. Enemy information can be rapidly formatted via an automated report. This information is forwarded to all other FBCB2 users and the ASAS supporting the user, usually the TF or brigade S-2.

B-139. FBCB2 supports the call-for-fire process via a message in JVMF sent directly to AFATDS. The integration of the laser ranger finder with FBCB2's ground positioning system greatly improves speed and accuracy of both calls for fire and enemy SPOTREPs. It provides key information to the CSSCS on unit logistics status.

Tactical Airspace Integration System

B-140. TAIS is a digitized, integrated airspace management and decision support system assisting the ground commander's role in the air battle. It supports automated AC2 planning and operations and ATS. It also helps planners build Army input for the joint ACO to distribute the approved AC2 overlay. TAIS can display ACMs in two or three dimensions while monitoring the real-time airspace situation. TAIS provides SU of the third dimension by providing real-time airspace information that displays location and movement of aircraft transiting the OE overlaid against current ACMs.

B-141. A TAIS is found at the division main CP. A second TAIS is located at the division tactical CP or aviation brigade where it can optimally provide flight-following functionality. At corps level, one TAIS is at the main CP while a second is placed consistent with the tactical situation. TAIS is also at theater.

Capabilities

B-142. TAIS deconflicts (mathematically and graphically), in real time, airspace usage in the third and fourth dimensions (altitude and time). For example, the operator can graphically rotate a three-dimensional representation of the airspace to see ACMs from different angles, enabling him or her to see how they intersect and overlap.

B-143. The ATS display includes information from the ACO and ATO. TAIS operators can use this display to track aircraft flight. If an aircraft leaves the safe transition corridor, TAIS can alert the operator. TAIS is able to communicate (voice and data) with current and future military aircraft (joint/combined), civilian aircraft and air traffic control systems and other U.S. and allied forces airspace users.

Digital Topographical Support System

B-144. DTSS enables topographic support personnel to receive, format/reformat, store, retrieve, create, update, and manipulate digital topographic data. It gives digital terrain analysis, terrain databases, updated terrain products, and hard-copy reproduction of topographic products to include maps. Its tactical decision aids support COA analysis and the MDMP. These aids include mobility analysis, intervisibility (LOS) analysis, environmental and climatology analysis, terrain elevation, and other special products. Using the global broadcast service, DTSS receives and distributes digital terrain data from the NGA. DTSS can update existing digital maps from satellite imagery and produce full-size, color paper maps from any DTSS product. DTSS is found at the corps main CP and tactical and brigade CPs.

Capabilities

B-145. DTSS produces sophisticated mobility analysis products. For example, it provides a detailed analysis comparing off-road mobility of the HMMWV and M1 tank.

B-146. DTSS performs intervisibility analysis, which is overlaid on a terrain map backdrop. For example, from any point on the map, it can depict every other point within LOS of the first point.

B-147. DTSS depicts a three-dimensional view such as a fly-through area. Colored areas show threat and friendly AD domes superimposed on satellite imagery. The DTSS database contains detailed terrain information but not weapon characteristics and locations; these must be obtained from the intelligence staff.

INTEGRATED METEOROLOGICAL SYSTEM

B-148. IMETS is the meteorological component of ABCS. It provides an automated, high-resolution weather system to receive, process, and disseminate current weather observations, forecasts, and weather and environmental effects decision aids. IMETS workstations, manned by staff weather teams, are at the aviation brigade, division, and corps main CPs.

Capabilities

B-149. IMETS receives and integrates weather information from polar-orbiting civilian and military meteorological satellites, the Global Weather Center, artillery meteorological teams, remote sensors, and civilian forecast centers.

B-150. IMETS processes and collates forecasts, observations, and climatological data to produce timely and accurate weather products tailored to the warfighter's specific needs. Additional weather information is available via the IMETS web pages. Severe weather warnings are disseminated to units via U.S. message text format message.

B-151. The integrated weather effects decision aid displays weather effects on weapon systems or missions. For example, it can show various weather effects–whether favorable, marginal, or unfavorable–on various weapons over the next 24 hours.

ARMY TACTICAL COMMAND AND CONTROL SYSTEM

B-152. ATCSS is a family of automated C2 tools. ATCCS consists of the following:

- MCS.
- Maneuver control system-light (MCS-L).
- ASAS.
- All source analysis system-light (ASAS-L).
- AFATDS.
- AMDWS.
- CSSCS.

Maneuver Control System

B-153. MCS is the S-3's tool. It displays the current battle and enables planning for the future battle. It provides the ability to collect, coordinate, and act on near-real time battlefield information. MCS integrates information horizontally and vertically to provide the COP with friendly, enemy, and noncombatant locations. MCS is found at echelons from battalion through corps.

Capabilities

B-154. A message processor is available on all MCS workstations. It is used to create, edit, transmit, print, and store messages in both U.S. message text format and JVMF.

B-155. With word-processing templates and web-browser technology, MCS can rapidly produce and distribute OPLANs, OPORDs, FRAGOs, and WARNOs. Task organizations are created, edited, and displayed using the unit task order (UTO) tool.

B-156. MCS collaborative planning tools enable multinode collaborative planning sessions within or between CPs. These tools include data conferencing, chat, and whiteboard. The whiteboard is a powerful capability for war-gaming, orders briefs, and back-briefs. The chat feature is similar to current chat programs available on personal computers. Multiple users can communicate simultaneously by posting text messages that can be read simultaneously by all chat participants.

Maneuver Control System-Light

B-157. MCS-L operates as a client of Medical Support Command. It is able to obtain data directly from the JCDB and update it with friendly locations and battlefield geometry. The main difference between MCS-L and MCS is the ability of the latter system to perform various net server functions and interface with FBCB2. MCS-L is found at battalion, brigade, and certain separate companies.

Capabilities

B-158. The MCS-L can be used to—

- Produce orders, plans, and annexes; used to develop task organizations, overlays, and synchronization matrices.
- Develop and assess COAs; MCS-L includes a distance/rate tool.
- Create messages and generate reports; used to maintain the staff journal.
- Record and depict NAI, target areas of interests, and CCIR including HVTs and HPTs.
- Function as file transfer protocol client/server; MCS-L possesses Adobe Acrobat, a file zip utility, Microsoft Office, and web browser.

All Source Analysis System

B-159. ASAS is the intelligence fusion system. It receives and processes intelligence and information from sensors, processors, and communications systems at national, theater, and tactical echelons and SPOTREPs from FBCB2. It provides a timely, accurate picture of the enemy situation. The S-2 uses his ASAS remote workstation (RWS) for automated situation development, COAs, targeting, tactical warning, and BDA.

B-160. ASAS is at echelons from battalion to corps. An ASAS RWS can function as a stand-alone system or an adjunct to an analysis and control element (ACE) at corps and division level and the analysis and control team at brigade level.

Capabilities

B-161. Intelligence personnel can use the analysis tools in the ASAS RWS for their IPB. For example, it is able to depict tracked vehicle GO and NO-GO areas overlaid on a terrain map. The ASAS RWS assists the warfighter's COA analysis with information on enemy units, equipment, locations, and movements.

B-162. Using reports and sensor inputs, the RWS can alert operator to enemy targets and automatically nominate them for friendly supporting fires. Commanders and staff can even focus ASAS on specific types of targets best supporting the mission.

B-163. ASAS also monitors the current enemy situation. Using the latest combat information and intelligence, it maintains and displays timely, detailed data on enemy units.

All Source Analysis System-Light

B-164. ASAS-L has vertical and horizontal interoperability with MCS, AFATDS, FBCB2, and other ASAS terminals. It is intended primarily for those who use preprocessed intelligence information and graphic IPB products from the analysis and control team, ACE, and the S-2's ASAS RWS (chief ASAS platform at corps, division, and BCT echelons). ASAS-L receives and processes initial INTREP and information received via FBCB2. It forwards these reports to the analysis and control team and ACE where the information undergoes intelligence processing and integration before returning to the brigade S-2 as fully correlated intelligence information. ASAS-L is located at battalion.

Capabilities

B-165. The ASAS-L provides ISR management and analytic support to the battalion S-2 for SU, tactical warning, force protection, and targeting. It provides an analyzed enemy picture to the operational picture.

Advanced Field Artillery Tactical Data System

B-166. AFATDS is the artillery management system employed by FS personnel. It provides fully automated FS planning, coordination, and control of close support; counterfire; interdiction; suppression of enemy ADs; and operations in unassigned areas. AFATDS matches FS weapons with targets based on target type, commander's guidance, unit availability, weapon status, and ammunition availability. It encompasses FS platforms across the services–including mortars, FA cannons, rockets, missiles, CAS, attack reconnaissance helicopters, and naval surface fire support (NSFS). AFATDS is a multiservice system.

B-167. AFATDS is at the firing platoon through theater Remote terminals allow commanders, LNOs, and other FS personnel to monitor FS operations and issue guidance.

Capabilities

B-168. AFATDS analyzes a potential target and then identifies which available FS systems are most effective. This information is shown to the operator through a visual display.

B-169. Based on command guidance, AFATDS prioritizes targets and supported units, specifying the method of engagement and volume of fire for each type of target. These priorities can vary according to specific guidance for each phase of an operation to best support the commander's intent and scheme of maneuver.

B-170. AFATDS processes fire missions through combat messages in dialogue with MCS, CSSCS, AMDWS, and FBCB2 and reports mission results to ASAS.

B-171. In addition to managing the FS of current operations, AFATDS assists FS planning for future operations. Its planning mode offers decision aids and analytical tools to determine which FS plan best supports a COA.

Air and Missile Defense Work Station

B-172. AMDWS is the AD system that enables monitoring of the current air operation while planning for future events. It also provides SU of the third dimension. The force operations capability of AMDWS supports the planning, coordination, and preparation for and sustainment of the AD mission. It integrates AD fire units, sensors, and C2 centers into a coherent system for defeating the aerial threat. Defense planning and analysis functions support development of AD missions and distribution and merging of missions between echelons. AMDWS also supports air battle management by displays that show ACOs, current fire unit status, alert posture, missile expenditure, and personnel ready for duty. AMDWS is located at the AD battery CP with the BCT main CP, division CPs, corps CPs, and theater.

Capabilities

B-173. The AD unit status screen shows location, alert status, on-hand munitions, vehicles, and personnel for AD units from section through battalion echelon.

B-174. Its weapon and sensor visibility feature supports placement of AD weapons and sensors. By analyzing platform capabilities and digitized terrain elevation data, AMDWS can determine area coverage of weapons and sensors at different locations.

B-175. The AMDWS mission planner shows zones of sensor coverage, weapons coverage, friendly and hostile air tracks, air avenues of approach, and airfields. The commander can use this display to synchronize AD coverage with the planned scheme of maneuver. Operators can set parameters to depict aircraft at various altitudes based on the surrounding terrain.

Combat Service Support Control System

B-176. CSSCS is the automated system for planning and controlling the sustainment of combat operations. Warfighters can logistically assess future COAs using current or planned task organizations and approved planning factors. CSSCS tracks the maneuver sustainment posture throughout task organization down to company level. CSSCS terminals are found from the battalion through theater.

Capabilities

B-177. Logistics reports depict unit and resource status with a color code of green, amber, red, or black by using corresponding percentages set by the user. Reports can be displayed as web-based custom reports or standard, preformatted reports. The standard report shows the logistics readiness of a unit and its subordinate units. The user can focus on parts of the report to isolate specific units and materiel items. This capability helps identify how an individual status affects the overall readiness rating of the unit. In the custom report, the user can track the status of specific units and resources.

- The capability report shows a unit's logistics ability to conduct sustained combat operations; this report provides unit resource status in relation to combat posture and intensity for the current day and next 4 days.
- The supply class report shows resource status with items grouped by class of supply.
- The personnel daily summary depicts unit personnel status and is available for all company-size units and separate battalions.

Command Post of the Future

B-178. Another collaboration tool being developed is the CP of the future. It is a suite of executive level decision support systems providing SU and collaboration tools to support decisionmaking.

B-179. It is designed to support parallel, synchronous and asynchronous, cross functional planning and execution. Team members share workspaces that embody their thinking about the current situation and collaborate to create a rich, multi-perspective, shared operational picture.

B-180. CP of the future operates over MSE/local LAN and will eventually be used over SATCOM.

B-181. CP of the future is not a replacement for the ABCS, rather it depends on ABCS for the majority of its data. It is partially integrated with the ABCS and limited to receiving one-way feeds from MCS, FBCB2, ASAS, and MTS. However, the Defense Advanced Research Projects Agency is working on 2-way ABCS integration and adding air and fire pictures.

B-182. CP of the future creates a more user friendly environment with which to manipulate the data and conduct collaboration. The CP of the future consists of the four main applications shown in figure B-13, page B-32.

B-183. The workspace consists of a two-dimensional map with a personal pasteboard for the commander and shared pasteboard for other participants, three-dimensional map, and VoIP. The commander's pasteboard provides the ability to draw and highlight data, force tracking, and map imagery. The pasteboard for other commanders and staffs contains real-time updating, master schedule; SITREP table; size, activity, location, unit, time, and equipment report table; and other displays as desired. All of the displays are shareable.

B-184. CoMotion client application provides a versatile commander's view into geospatial, temporal, and other forms of data.

B-185. The Oculus map provides a three-dimensional picture which allows terrain manipulation and rapid terrain appreciation. It can be linked to any CoMotion client map, showing same data and area. It can provide entity representation of BLUE elements. The time slider function allows visualization of forces flowing over time.

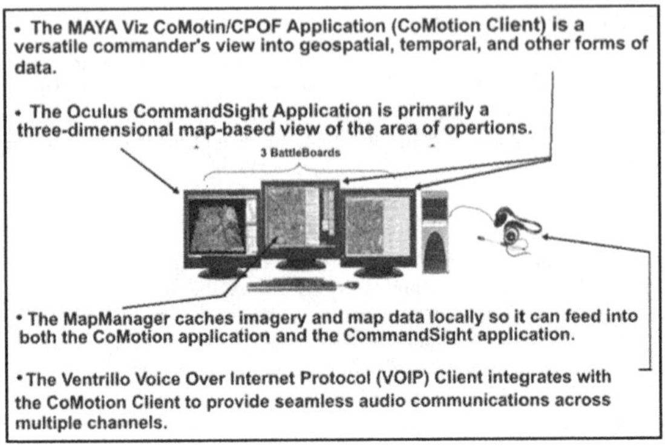

Figure B-13. Client system—four main applications

B-186. The MapManager application caches imagery and map data locally to be utilized on the CP of the future, and the VoIP application integrates with the CoMotion client application to provide seamless audio communications across multiple channels.

DIGITAL COMMAND POST OPERATIONS

B-187. The Army is making rapid and drastic changes in CP design, taking full advantage of the newest computer technology. The CPs for digitized units will be mobile, deployable, and equipped to access, process, and distribute information and orders for their echelon. This section outlines the internal operations of a digital CP. FM 71-100-2, FM 71-100-3, and FM 5-0 contain detailed discussion.

DATA EXCHANGE

B-188. Central to digital CP operations is the manner in which they exchange data. ABCSs share information either directly with one another or through the JCDB. The JCDB resides on all ABCS computers in a CP and provides data for common applications that generate the COP. Battlefield information dynamically flows back and forth between ABCSs and the JCDB. When data are entered through a BAS, this change is forwarded to all ABCS subscribers on the CP's TACLAN and posted to the COP (figure B-14).

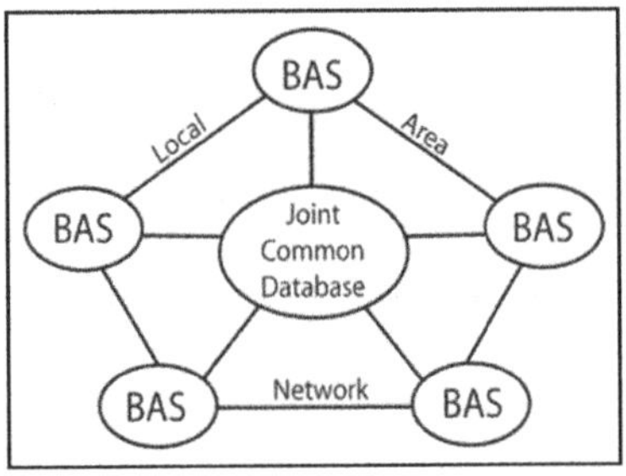

Figure B-14. Data exchange within a command post local area network

B-189. Data are also exchanged between CPs. This exchange allows the same data to be maintained in the JCDBs in different CPs. Data generated by each BAS flows to its counterpart BAS at adjacent echelons. Each BAS then transfers this information to the JCDB at that echelon via the TI. Friendly position information flows from FBCB2 upward through the server located at each echelon. This information is then deposited into that echelon's JCDB. This data exchange ensure all CPs have JCDBs resembling one another. This is key to creating the COP. Figure B-15 shows data flow between an example battalion and brigade with their MCS operating as servers. Note the flow of friendly position information (depicted by dashed arrows) moving between these echelons and into their JCDBs. Each BAS can, in turn, access this friendly picture from the JCDB at its echelon. The flow of data from a BAS to other BASs and the JCDB is shown by solid arrows.

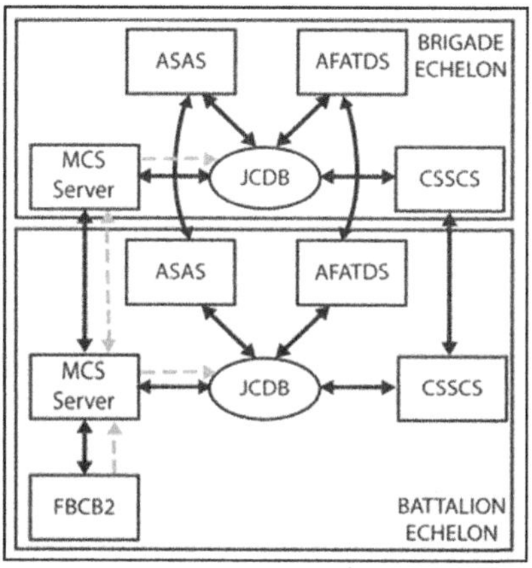

Figure B-15. Example of a data exchange between command posts

DIGITAL COMMAND POST LAYOUT

B-190. The standardized integrated command post system (SICPS) is the new generation of CP facility systems to support digitized units. SICPS is a C2 enabler, providing the platform to conduct digital CP activities. Its primary purpose is to support C2 of digitized units by housing their ABCSs. SICPS is designed to facilitate CP operations by providing flexibility, commonality, and operational capabilities needed to enhance unit mobility and integrate ABCS and associated communication and networking equipment. It supports integration of these command, control, communications, computers, ISR assets into platforms that can serve as a stand-alone CP or an integrated element in a larger digitized CP.

B-191. The SICPS has seven CP variants to include track- and wheeled-vehicle-mounted vans, tents, and hard shelters. The digital CP collocates staff sections and supporting communications systems to facilitate both face-to-face interaction and digital data exchange.

B-192. Unit SOPs will dictate the CP layout. As with the analog CP, the digital CP's physical setup must facilitate communication and analysis of information as well as accommodating computer hardware requirements. Within the digital CP, information is processed at two locations—individual workstations and the combat information center (CIC).

INDIVIDUAL WORKSTATIONS

B-193. The focus of the individual workstation is the individual BAS and specific Army WFF it supports. At his workstation, the staff member inputs and monitors data within his sphere of responsibility. He or she also accesses data posted to web pages and shared files by other staff sections in the LAN and WAN to carry out Army WFFs and duties.

COMBAT INFORMATION CENTER

B-194. The focus of the CIC is integrated battle monitoring and decisionmaking. It is a special location within the CP for the display of information. The CIC is the central area for viewing information for the commander and his staff to maintain SA.

Large Screen Display

B-195. The large screen display (LSD) is the only area in the CP where all key BAS data can be viewed simultaneously. It is the place where battlefield vision is best supported. The commander uses the CIC to illustrate his guidance and, with his staff's assistance, to develop and maintain the COP. CICs vary by MTOE. However, the typical CIC has two LSDs, each capable of displaying nine subscreens. Each subscreen can display the COP and be configured in various ways to best support the commander's information display preferences. The more subscreens used, the lower the image resolution. It is recommended each LSD screen use no more than four subscreens. Two LSDs allow display of eight subscreens. The addition of the engineer battalion LSD increases this display capability.

Data Display Management

B-196. IO plays a key role in a commander and staff's ability to maintain an accurate picture of the battlefield in the CIC. With feeds from each ABCS, the LSD enables them to see more of the battlefield and receive greater amounts of real-time battlefield information by Army WFF than is available with analog systems.

B-197. More information is not necessarily beneficial to mission planning and accomplishment. Data must be filtered, fused, and focused to create meaningful informational displays relevant to the mission. These displays or tactical pictures must be presented in a logical manner on the LSD to support SU. CP digitization has replaced analog maps, acetate, and wing-boards with digital overlays and electronic files. Because electronically stored information is readily available through a minimum number of computer keystrokes, there is less need to print paper copies of the information. Leaders and staff must know what data are available to them to make decisions about what will be displayed.

B-198. Although the LSD can display any BAS electronic data, the narrative and static aspects of some information still lend themselves to paper-copy posting within the CP. This is especially true for information less likely to change during a mission such as CCIR and the synchronization matrix. In turn, this optimizes the use of LSD subscreens by freeing them to depict dynamic ABCS digital content. The commander, XO, S-3, and battle captain must be able to orchestrate Army WFF coordination through the display of key information on the LSD. Each staff section must, therefore, maintain information relating to its Army WFF using visual graphics that support the COP. Staff sections and their supporting systems should be arranged around the LSD to facilitate information control, interaction, coordination, and information analysis.

B-199. The COP is displayed on the LSD through one ABCS, typically the S-3's MCS or MCS-L. COP control and manipulation and CP LAN administration are aided by centrally collocating the CP server and BAS that projects the COP. The ability to view the LSD through the BAS controlling the COP also facilitates communication and navigation through data. During discussions in the CIC, personnel can focus staff on key portions of the COP. Data are displayed on the LSD via the COP using the ABCS COP application or through overlays provided by individual BASs. To portray the COP graphically requires METT-TC analysis of information. The COP displays enemy (red feed and graphics), friendly (blue feed and graphics), terrain (characteristics and impact), and civilian considerations (gray feed and graphics).

B-200. Friendly analysis occurs in the CIC by all sections and systems. Each BAS provides Army WFF overlays for subsequent data manipulation and consolidated viewing of operational pictures that form the COP. Enemy analysis is especially time-sensitive information.

B-201. The MCS whiteboard or electronic whiteboard equips leaders and staffs to conduct collaborative sessions. Participants at distributed locations view the same enemy and friendly COP on an MCS display and are linked with audio. The telestration feature of whiteboard allows each participant to use a mouse with a crayon drawing capability to visually depict locations, graphics, and other coordination measures seen on the participants' screens.

DIGITAL RUNNING ESTIMATES

B-202. Not all key information can be graphically depicted on the LSD. Such information must be captured in a readily available, continuous update format for quick dissemination and assimilation. FM 5-0 emphasizes each staff section should maintain a running estimate (in narrative form, at division and higher and, in graphical form, at brigade and battalion). In the analog CP, these graphical running estimates correspond to the wing board and map data.

B-203. Digitization has eliminated the need to post information to wing boards but has created the need to organize digital data. Units must capitalize on the TACLAN web pages maintained by each staff section for organizing and posting critical mission data. By placing digital running estimates on a web page, each staff section supports the commander and staff needs to quickly review, update, and use information for battle monitoring and planning.

B-204. Establishing a standard running estimate format facilitates navigation through the estimate and cross-referencing between estimates. Running estimates should also list available Army WFF overlays by name to better focus graphical review within the ABCS COP application and to focus all echelons and staff on the same, most current data. Through digitally equipped LNOs, analog units should access these digital running estimates to obtain current operational data and help synchronize their operations with digital units.

INFORMATION MANAGEMENT

B-205. The staff must be organized to support the IM process of filter-fuse-focus. This process is guided by doctrine, TTP, and unit SOPs. The staff must operate according to established procedures that specify access to common databases, common displays, and report formats. The staff must be organized to allow vertical and horizontal flow of information. This organization should provide links between teams within staff sections, staff sections within a CP, and CPs at the same, higher, and lower echelons.

B-206. Digitization enables commanders and staff members to focus more on execution of combat operations and much less on planning, coordination, and processing of information. Commanders and staff will have much more data upon which to base their decisions. Their challenge will be to manage the flow of vast amounts of data so the right information gets to the right person at the right time. These specific challenges are—

- Relevancy: Determine relevant information from among the vast amount of data available.
- Responsibility: Ensure each product is the assigned responsibility of a specific staff section.
- Accuracy and currency: Ensure data are correct and up-to-date.
- Dissemination: Ensure information generated by the staff is supplied to the right personnel.
- Evaluation: Ensure information is appropriately assessed.

Relevancy

B-207. Because of the large quantity of data available, the commander needs to establish information priorities to focus the staff during data collection. These priorities must address information relevant to the specific operation. The commander provides this focus via CCIR that are—

- Specified by the commander and applicable only to him or her.
- Situation dependent and linked to present and future operations.

- Based on predictable events or activities.
- Time sensitive (answers to CCIR must be reported to the commander by the most rapid and effective means).

B-208. Table B-4 summarizes the CCIR responsibilities.

Table B-4. Commander's critical information requirement responsibilities

Duty Position	*Sample Briefing Items*
Commander	Establish CCIR. Establish priorities for information collection and distribution. Assign assets to collection of information. Determine display of information throughout his command during an operation.
Chief of Staff/ Executive Officer	Manage CCIR. Establish TTP for tracking when and how CCIR are answered. Assign responsibilities to personnel within the staff sections and CPs to manage information. Supervise commander's guidance for collecting, processing, and circulating information.
Staff Leaders	Manage information within WFFs. Recommend CCIR based on analyses. Record, evaluate, analyze, and report collected information to answer CCIR.
Staff Section Operators	Monitor ABCS traffic. Know what to file, what data to display, what to name/rename files, and where to store them. Know what graphics to display. Be alert to CCIR and know how to act on CCIR for these requirements.

Accuracy and Currency

B-209. ABCS is automated allowing information to flow quickly and accurately; however, most of its information does not flow automatically. Only friendly position data, which supports the friendly or blue picture, flows automatically via FBCB2 and the TI. For all other data to enter and flow throughout ABCS each BAS must be properly initialized and its data maintained. Staff sections will have ready and routine access to many products of other staffs and units at varied echelons. This outside access may take place without a staff section's knowledge. Staffs must ensure they continuously post their most up-to-date products and maintain them on staff web pages or shared folders. CP internal procedures must specify routines and suspenses for producing and revising ABCS products and specify where they will be maintained.

Dissemination

B-210. Because of bandwidth limitations, it may not be possible to routinely send products through e-mail; but, it is not enough to merely post information to a web site or shared folder and expect others to use it. With the exception of routine, scheduled postings and updates, the staff must proactively notify users when such changes are made. When a product is posted or revised, staff sections must notify other staff sections and units at the same, lower, and higher echelons. This notification must include instructions on precisely where to find the product and its file name. Units must establish SOPs that specify file-naming conventions and file-management procedures. Whether forwarding products or providing notification of product postings in shared files/web pages, the right personnel must receive the right information. Correct address information using the ABCS address books and message handling tables must be established to ensure data are sent to the correct BASs. Addressees must be the users employing the individual ABCS rather than generic role names in the address book. If this is not done correctly, information on one BAS will not flow to other BASs even in the same CP. During initialization, operators must also create and distribute databases, which can be done via messages in ABCS. These databases ensure BASs can share correct information.

Evaluation

B-211. Computer data tends to be accepted at face value as it is computer-based and assumed to always be correct. Users of digital systems must resist this tendency. Error can be introduced through failures in BASs, databases, and communications systems; human error in inputting data; and failing to update information in a timely manner. Data must be evaluated within the context provided by SU to verify that they are accurate and current. Users must follow up on discrepancies to ensure they have the correct information.

DIGITAL DUTIES AND RESPONSIBILITIES

B-212. The diverse products produced using ABCS must be the responsibility of specific staff sections. This responsibility is usually obvious, being based on doctrine. Unit SOPs/TTP must confirm these doctrinal responsibilities while ensuring all other products are the assigned responsibilities of specific staff sections.

B-213. Staff functions as described in FM 5-0 will not fundamentally change in the digital CP. However, these functions will be carried out differently using the digital tools ABCS provides. Digitization also requires personnel to perform new functions as listed below. These digital CP tasks should be conducted in addition to, and as a part of, standard staff responsibilities.

COMMANDER

B-214. The commander has the following digital duties and responsibilities:

- Provides command guidance for employing ABCS.
- Provides C2 of automation resources.
- Establishes automation support priorities.
- Specifies unit COP.
- Establishes the CCIR and ensures these requirements are depicted in ABCS.
- Ensures subordinate leaders are trained in employment, operation, and sustainment of automation.
- Trains subordinate leaders and staff to create, maintain, distribute, and use the COP.

EXECUTIVE OFFICER

B-215. The XO has the following digital duties and responsibilities:

- Coordinates the staff to ensure ABCS integration across BAS.
- Ensures staff integrates and coordinates its ABCS activities internally, vertically (with higher headquarters and subordinate units), and horizontally (with adjacent units).
- Manages the CCIR; ensures satisfaction of the CCIR.
- Directs creation and distribution of the COP to include procedures for updating enemy and friendly SU.
- Monitors information filters, collection plans, and networks that distribute the COP.
- Provides guidance for automation support.
- Coordinates the staff to ensure automation support.
- Coordinates procedures for inter-CP VTCs and whiteboard sessions.
- Monitors liaison teams with analog (nondigitized) units and joint/allied forces for their contribution to the COP.

Personnel Officer

B-216. The S-1 has the following digital duties and responsibilities:

- Is responsible for personnel functions of CSSCS.
- Employs CSSCS to monitor and report on personnel-related portions of the commander's tracked item list.
- Manages Electronic Military Personnel Office (enlisted)/Total Officer Personnel Management Information System II (officer) interface with CSSCS.

Intelligence Officer

B-217. The S-2 has the following digital duties and responsibilities:

- Acts as staff proponent for ASAS and IMETS.
- Supervises ASAS and IMETS operations and support.
- Provides guidance on employment and support of ASAS and IMETS.
- Supervises the information security program; evaluates security vulnerabilities.
- Assists the S-6 in implementing and enforcing LAN security policies.
- Provides software application expertise on proponent systems.

Operations Staff Officer

B-218. The S-3 has the following digital duties and responsibilities:

- Acts as staff proponent for MCS, AFATDS, AMDWS, FBCB2, and AMPS.
- Plans, integrates, and employs ABCS.
- Develops the ABCS annex for plans and orders.
- Develops ABCS annexes to garrison and TACSOPs.
- Oversees IO.
- Provides operational and support guidance regarding network employment to subordinate units.
- Integrates AMPS and distributed planning data.
- Creates, maintains, and displays the COP; maintains SU of all units.
- Coordinates with S-6 for communications connectivity in support of ABCS.
- Plans and monitors operator digital sustainment training.
- Provides software application expertise on proponent systems.
- Assigns LNOs and coordinates digital support.
- Collects and distributes postmission results/BDA.

Logistics Staff Officer

B-219. The S-4 has the following digital duties and responsibilities:

- Acts as staff proponent for CSSCS.
- Supervises CSSCS operations and support.
- Provides guidance on employment and support of CSSCS.
- Monitors and reports on status of all automation equipment.
- Provides software application expertise on proponent systems.

Signal Officer

B-220. The S-6 has the following digital duties and responsibilities:

- Serves as signal SME to the commander; advises the commander and staff on all signal support matters.
- Monitors WAN performance; integrates the CP LAN.

- Is responsible for all automation information systems, automation and network management, and information security.
- Ensures consistency and compatibility of automation systems.
- Manages the TI; is responsible for network employment, configuration, and status monitoring and reporting.
- Receives planning worksheets with LAN/WAN requirements.
- Ensures unit information network connectivity between unit and higher/lower echelons.
- Plans, coordinates, and manages network terminals.
- Develops, modifies, and manages network need lines, UTO, and base configuration files.
- Plans, coordinates, and manages communications links to include reach-back communications.
- Coordinates with higher echelon S-6s for additional communications support.
- Develops and coordinates the signal digital support plan.
- Determines system and retransmission requirements for the tactical situation.
- Coordinates with higher, adjacent, and subordinate units in development of the signal digital support plan.
- Manages the release of ABCS software within the unit.
- Provides a focal point for automation support (help desk).
- Implements and enforces LAN security policies.
- Establishes COMSEC accountability, distribution, destruction, and security procedures within the unit.

Mission Application Administrator

B-221. The mission application administrator has the following digital duties and responsibilities:

- Assists the S-6 in managing the network.
- Plans and coordinates the linking of BAS to the unit CP.
- Supervises and performs unit-level maintenance and installs and performs maintenance on multifunctional/multi-user information processing systems, peripheral equipment, and associated devices in mobile and fixed facilities.
- Performs analyst functions; constructs, edits, and tests computer system programs.
- Performs preliminary tasks necessary for CP LAN initialization.
- Assists in troubleshooting digital systems.
- Conducts data system studies and prepares documentation and specifications for proposals.
- Maintains master copies of software.
- Backs up data for user-owned and -operated automation information systems.
- Assists in recovery of digital data at the user level.
- Operates and performs PMCS on assigned vehicles and power generators.
- Monitors BAS PMCS program.
- Coordinates repairs with the S-6 section.

Battle Captain/Battle Staff Noncommissioned Officer

B-222. The battle captain/battle staff NCO has the following digital duties and responsibilities:

- Oversees operations of assigned BAS.
- Controls/directs initialization of the BAS within the CP LAN (battle staff NCO).
- Ensures information flow and coordination take place between and within each staff section and with higher, adjacent, and lower headquarters.
- Accesses and employs information through ABCS in support of operations and planning.
- Ensures key BAS products are available and current in support of the mission.

BATTLEFIELD AUTOMATED SYSTEM OPERATORS

B-223. The BAS operator has the following digital duties and responsibilities:

- Installs and operates assigned digital hardware and software.
- Establishes connectivity of assigned BAS within LAN/WAN; ensures the system interfaces with correct tactical communications.
- Inputs operational data.
- Produces automated reports required by commanders and staff leaders.
- Performs PMCS on assigned BAS.
- Isolates, identifies, and tracks digital system problems.
- Maintains continuity of digital operations.
- Maintains portions of the COP, as assigned.
- Ensures unit-level information security.

MANAGEMENT OF DIGITAL COMMAND POST PERSONNEL

BATTLE ROSTERS

B-224. Each section within the CP must maintain a digital battle roster listing the section operators assigned to each BAS. At a minimum, sections should plan for three operators per system. Two Soldiers man a 12-hour shift each plus one Soldier serves as a backup and provides periodic relief. The roster should list the following:

- Personnel name and rank.
- Assigned BAS.
- Assigned shift.
- Date of most recent training on system.
- Software version of most recent training.
- Estimated date of departure from unit.

B-225. Operators are managed in a manner similar to unit vehicle drivers according to the following principles:

- Depth: Have more trained operators than needed to ensure BAS coverage even when unanticipated losses occur.
- Anticipate: Know when personnel are scheduled to depart the unit, and train their replacements well in advance.
- Leaders: Section leaders should be prepared to function as operators; in addition to providing additional coverage, this ability enables section leaders to better supervise and employ the BASs they oversee.
- Currency: Operators must be trained on the most current software carried on their BAS.

SHIFT MANAGEMENT

B-226. Shift changes are usually scheduled at 12-hour intervals. Commanders consider offsetting shift changes at midshift for key personnel. Staggering personnel in this manner maintains a constant interface of new and old shift personnel. This practice ensures at least one individual knows what happened during the previous shift. Figure B-16, page B-41, provides an example.

B-227. Soldiers must conduct a one-on-one exchange of information with the person they are relieving. This exchange must be followed by section wide debriefs to ensure continuity in information flow and handoff of ongoing staff actions.

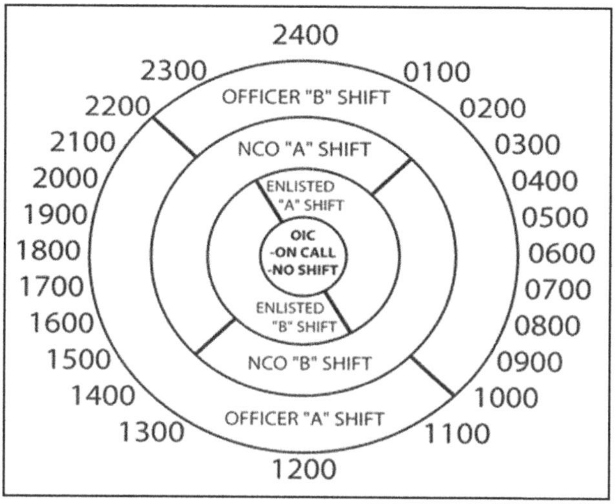

Figure B-16. Example of staggered shift changes

B-228. Following the individual brief, section-level products and actions are reviewed. Each staff section accomplishes the following actions:

- Review digital journal for the past 12 hours.
- Review and update any CCIR.
- Review current approved overlays.
- Review current COP products.
- Check files to ensure standard naming conventions are used.
- Review the UTO.
- Check section web products for updating and ensure they are properly posted.

B-229. A collective information exchange, in the form of a shift change brief, must be conducted so the incoming shift receives a positive change of control. Personnel from different staff sections will have access to key information produced by other sections and CPs. Handover briefings focus much less on the rote exchange of information. Rather, these briefing sessions can function to focus personnel on available information, evaluation of information, status of current operations, and tasks supporting future operations.

B-230. Critical digital considerations are briefed collectively within the CP. Table B-5, page B-42, provides an example of what this brief may look like. There is presently no doctrinal guidance on this process. Units should develop SOPs to address this requirement.

Table B-5. Example of a command post shift change brief

Staff Position	*Sample Briefing Items*
S-3 Battle Captain	Current higher and brigade changes to task organization. Disposition /status of units. Current and future missions. Current operations and timelines. LNO updates. Combat power status. Projected operations over next 12 hours.
S-3/Air	Brief current and future AC2. Provide AC2 status in conjunction with AD office, FS officer, and ALO.
S-2/Weather	PIR/CCIR. Provide SU and location/status of all ISR assets (national to division/brigade). Request for information to higher headquarters (ARFOR or national). Weather-next 12 hours impact/effects on friendly and enemy systems. HVT/HPT. BDA. Significant activities during past 12 hours.
FSE	Organization for combat. Unit locations and status. Priority of fires. HPT/attack guidance matrix. FSCMs. Significant activities.
ALO	Preplanned request status. Immediate request status. In-flight reports.
AD	Organization for combat. Current AD warning and control status. Aircraft engagements. Location and status of AD units.
Engineer	Operations since last update. Status of equipment and class IV/V. Future engineer operations. Recommendations for the commander.
Chemical	CBRN condition. Current and recommended MOPP. Enemy CBRN activity. Chemical unit locations and status.
S-1/S-4/Surgeon	Equipment status. Class VIII status. Priority of support. Personnel status/health service status.

BATTLE RHYTHM

B-231. Battle rhythm is a doctrinal term describing a process essential to effective and efficient battle staff operations. The cycle of recurring events within a CP focuses staff members to meet information and action requirements. These recurring events include—

- Shift changes.
- Targeting meetings.
- Reports.
- Battle updates without the commander.
- Battle update briefings.
- Commanders' collaborative sessions.
- Battle captain collaborative sessions.

B-232. The staff must achieve a battle rhythm for updating and viewing information and understand how to use it to effect operations. A well-established battle rhythm aids the commander and staff with CP organization, IM and display, decisionmaking, and fighting the battle from the CIC and via satellite C2 systems. Battle rhythm demands careful planning and design. The many competing demands must be deconflicted. Even subordinate units affect a higher echelon's battle rhythm based on their needs and unit procedures. Two key things to consider when establishing SOPs for battle rhythm are scheduled updates (both with higher and subordinate units) and bandwidth. ABCS competes for bandwidth with the commander's digital updates or VTCs primarily if data passes over communications links between CPs. The MDMP can have one of the most dramatic effects on battle rhythm. The process is lengthy and detailed and must be closely coordinated with other ongoing actions.

BATTLE UPDATE BRIEFING

B-233. The battle update briefing provides the commander with analyzed information essential to decisionmaking and to synchronize the staff's actions. COP use expedites the battle update and makes it more current. The more information used from the COP, the more time the staff has to analyze and evaluate the information. The battle update briefing itself centers on the COP displayed in the CIC. The staff must be selective as to what other information is presented given the wealth of data and availability at each BAS. Unit SOPs, command guidance, and operational requirements guide what information is briefed. Facts and capabilities may be presented in digital staff estimates for the commander to review before the briefing. This allows the battle update briefing to focus on by-exception information and specific commander issues. Methods to update the commander depend on his location, connectivity, and information he or she requires. Table B-6 compares delivery methods.

Table B-6. Update delivery comparison

Commander in an aviation CP	*Commander in another CP*
Verbal.	Voice (radio, phone).
Over the shoulder of an operator.	FBCB2.
Commander's update page and pull-up information.	MCS or access to another BAS at this location.
Links to staff section pages and pull-up information.	
Collaboration session.	Collaboration session.
LSD.	

B-234. Traditionally, these updates were a recounting of significant events since the last update. To build the update, the CP would establish an information cut-off time. The focus was on maintaining SU. ABCS has altered this briefing from a staff brief to a constantly available information package focusing on the commander's needs. Table B-7 shows how the briefing has evolved from its traditional analog form to its digital form.

Table B-7. Traditional versus digital

Traditional	*Digital*
Significant events since last update.	Commander accesses his own critical information needs.
Current as of cut-off time.	Updated continuously.
Periodic event.	Available anytime.
Current SU.	Enhances SU.
Staff presentations and their preparation were significant events.	Staff routinely maintains information files, which continues with normal operations.

B-235. Battle update briefs should maximize the use of information from BASs to aid in understanding the COP. Cutting and pasting information to non-ABCS briefing slides focus on fact finding and less on analysis. The traditional form also consumes considerable time–more than 1 hour to build/transmit slides, 1 hour to present (at brigade level), and 1 additional hour to present (at the division level). When slides are briefed, their information is outdated and inconsistent with the more current COP.

Analog Unit Interaction

B-236. Digitized units must be prepared to operate with nondigital units not having the technology to access the digital COP. Liaison parties will almost nearly be necessary to ensure full exchange of information between digitized and nondigitized units. The primary tasks of digital l teams are—

- Receipt and transmission of orders, graphics, and intelligence data via BAS.
- Provision of friendly and enemy SU to the analog unit using its BAS.
- Manual creation of the analog unit friendly and enemy SU and its transmission back to the parent organization.
- FS and coordination.

Planning

B-237. A digitized unit must exchange liaison teams with nondigitized units early and consistently throughout the planning process. Nondigitized units must strive to conduct parallel planning but will be at a disadvantage without digital staff tools. Parallel planning requires rapid exchange of information with analog units during the planning process. Involving higher, adjacent, and lower staff elements early in the planning process allows the entire staff to see both current and future operations and identify known or potential problem areas.

Liaison Teams

B-238. Digital liaison teams may be sent to the analog unit's CP. Liaisons provide at least some digital capability to analog units. These teams support SU for both the digital and nondigital unit, issue of orders, and informal information exchange. The number of liaison teams is limited, and these alone cannot solve the C2 challenges of analog units without digitally based SU. Liaison teams may be needed to escort elements of the analog unit, even down to single vehicles if necessary. This latter option provides SU for these analog elements but is only practical if the digital unit forms additional liaison elements.

Equipment Requirements

B-239. The equipment and skills required of the liaison teams are a function of the type of operation being conducted and force with which the team is coordinating. The following three basic forms of liaison effect task organization of liaison teams:

- **Digital unit to digital unit**: Requires the least equipment and personnel; information is easily shared in near-real time; critical SU is maintained in each unit's knowledge base.
- **Digital unit to analog unit**: May occur when conducting operations with some active component units, most RC units, and coalition forces; these teams require a full suite of digital systems to maintain the parent unit's COP and provide SU of the nondigitized force back to the digital headquarters. Representation from each staff section may be required on the team.
- **Digital unit to nonmilitary forces/agencies**: Is the same as for analog units but augmented with additional specialties such as the S-9/Assistant Chief of Staff-Civil Military Operations.

Appendix C

Personnel Recovery Operations

Even though PR can be performed by ground units, this appendix addresses use of aviation assets to affect a recovery. Army aviation PR is a mission performed by a designated aircrew for the specific purpose of PR when the tactical situation precludes SAR assets from responding, and survivors and their location have been confirmed. Mission success largely depends on thorough premission planning, accurate and timely intelligence, verifiable survivor location, flexible and redundant C2, and highly trained PR forces.

SECTION I – GENERAL

C-1. The Army's PR philosophy is one of leadership and accountability. It comprises primarily the Soldier's Creed, directed responsibilities, and practical considerations. The Army conducts PR as a collection of architecture and activities designed to affect the recovery of personnel who are isolated, missing, detained, or captured (IMDC). PR is no longer just CSAR, special operation force or air asset centric operation designed primarily for the rescue of aviators.

C-2. The Army PR function is defined as "the sum of military, diplomatic, and civil efforts to affect the recovery and return of U.S. military, DOD civilians and contractor personnel, and/or other personnel as determined by the Secretary of Defense, who are IMDC in an OE." PR is one of the highest priorities within the DOD. FM 3-50.1 outlines how the Army conducts PR within the joint services construct.

C-3. Army aviation's role is in the execution of pre-established procedures and well rehearsed operations to report, locate, support, recover, and repatriate IMDC personnel. While it is every Soldier's responsibility to assist IMDC personnel, aviation assets are generally employed in the recovery of personnel within the unit or supported units' AO when the IMDC personnel's location is known. Four principle methods of recovery are used when planning and executing recoveries—immediate, deliberate, external supported and unassisted.

IMMEDIATE

C-4. Immediate recovery is the sum of actions conducted to locate and recover IMDC personnel by forces directly observing the isolating event or, through the reporting process, determining that IMDC personnel are close enough to conduct a rapid recovery. Immediate recovery assumes that the tactical situation permits a recovery using the forces at hand without detailed planning or coordination.

DELIBERATE

C-5. Deliberate recovery is the sum of actions conducted by friendly forces when an incident is reported, and an immediate recovery is not feasible or successful. Weather, enemy actions, IMDC personnel location, and recovery force capabilities are examples of factors that may require the detailed planning and coordination of a deliberate recovery.

EXTERNAL SUPPORTED

C-6. External support recovery is the sum of actions conducted when immediate or deliberate recovery is not feasible or successful. External support recovery is either the support provided by the Army to other joint TF components, interagency organizations, or multinational forces, or the support provided by these entities to the Army. CAS, ISR, and airborne C2 are examples of capabilities that may be required from different components to execute an external support recovery.

UNASSISTED

C-7. Unassisted recovery comprises actions taken by IMDC personnel to achieve their own recovery without outside assistance. An unassisted recovery typically involves an evasion effort by IMDC personnel to return to friendly forces, or a point where they can be recovered via another method. While the code of conduct requires IMDC personnel to make every effort to evade or escape, commanders must strive to recover these personnel utilizing one or a combination of the other methods.

SECTION II - PLANNING

C-8. The ability of the aviation brigade to successfully conduct PR is a function of proper MDMP, risk management, training and providing resources. PR integration into all brigade missions trains critical skills needed at every level to allow the safe return of friendly forces and deny the enemy an easy way to capitalize on mistakes in planning. The planning and training of the CAB encompasses everything from the identification of high-risk isolating events for ground and air missions, to the execution of movement with gun escort to rapidly recover BCT Soldiers stalled with vehicle problems in a potentially unsafe area. It extends to supporting external joint assets with CAB assets to affect a recovery. In-depth planning guidance may be found in the FM 3-50.1. Some basic elements to consider are—

- The primary mission continues parallel to the recovery effort.
- The goal is recovery of the IMDC person.
  - Plan a system that enforces accurate reporting.
  - Ensure the system provides for accurate record keeping without degrading the PR effort.
- Prevent the IMDC event.
- Prepare for the IMDC event.
- Design the PR architecture within the C2 system.
- Organize, train, and equip for PR.
- Integrate contractor and DA civilians into OPLAN/OPORD.
- Transition the PR capability (IMDC events may occur early in the deployment).
- Plan for integrated rehearsals.
- Employ the recovery force based on METT-TC.
- Develop the plan from receipt of the report to reintegration of the IMDC.

C-9. The integration of the entire staff is crucial to developing sound PR plans during MDMP. Commanders ensure that PR is included in the MDMP, and in the eventual plan or order that results. PR planners coordinate their actions with all functional staff elements, leveraging the expertise of staff members in their individual areas to develop PR COAs and plans that enable PR coverage across the AO. By focusing planning on the five PR execution tasks (report, locate, support, recover, and reintegrate), and the abilities of commanders, staffs, units, and potential IMDC personnel to perform together to execute those tasks during a PR mission, PR planners develop robust PR plans that support the overall mission.

TRAINING

C-10. Training must be conducted at all levels within the brigade and constantly reinforced to be effective. PR academics and survival, evasion, resistance, and escape (SERE) skills allow each Soldier to understand how to survive an isolating incident and that there is a plan for their safe recovery. The brigade staff, with the personnel recovery officer (PRO), should develop plans and actions within CP drills and SOPs to facilitate the PR execution tasks. These elements should be integrated into garrison and combat daily operations. This may be accomplished by acting on routine precautionary landings and vehicle maintenance problems as if in combat, or adding isolating events to collective training exercises. Honest assessment of daily reactions to isolating incidents will give commanders a measure of effectiveness in their PR programs.

ROLES AND RESPONSIBILITIES

C-11. Each PR event has the possibility of becoming a joint mission depending on the situation of forces involved in a recovery. Some joint participants receive specialized training to execute their role in a recovery. A thorough understanding of the roles and responsibilities of all participants ensures recoveries that begin as immediate or deliberate may be continued as externally supported with a minimum of confusion. This level of functionality and modularity requires an understanding of terms, recovery training and action drill rehearsals at all levels (table C-1).

Table C-1. Personnel recovery terms

Joint	*Army*	*Civilian*
Joint Personnel Recovery Center (JPRC)	JPRC	Rescue Coordination Center
Personnel Recovery Coordination Cell (PRCC)	PRCC	Rescue Sub-center
PRO	PRO	SAR Mission Coordinator
On-scene Commander (OSC)	OSC	OSC
Airborne Mission Commander	S-3/Battle Captain/C2	Aircraft Coordinator
CSAR Unit	No Army term	SAR Unit
Helicopter Recovery Force	Helicopter Recovery Force	SAR Unit
Rescue Escort (RESCORT)	Gun Escort	No civilian term
RESCORT Commander	AMC (attack)	No civilian term

ON-SCENE COMMANDER

C-12. The OSC is the person designated to coordinate recovery operations within a specified area. He or she does not have to be in an aircraft; he or she may be ground or vessel based, but must be proficient in all PR procedures and have the ability to communicate with higher headquarters. While this qualifies a pilot to act as OSC, any Army aircrew may be called upon to act in this capacity. In fact, if any aircraft goes down, the first aircraft to arrive on scene (wingman) assumes OSC responsibilities regardless of proficiency. The OSC checklist may be found in the theater SPINS. Other responsibilities of the OSC include—

- Establishing and authenticating communication with isolated personnel.
- Locating isolated personnel and passing initial information to the AMC via the rescue mission brief.
- Conducting a threat assessment of the objective area (avoid highlighting the isolated personnel's location).
- Completing the OSC checklist.
- Determining the health/condition of isolated personnel and passing status to the AMC.
- Re-authenticating isolated personnel after OSC changeover only when the situation warrants.

RESCUE MISSION COMMANDER

C-13. The rescue mission commander (RMC) is the designated AMC maintaining control of the entire recovery during the launch, en route and terminal phases. Careful consideration of RMC selection should include knowledge of the overall mission, capabilities of the helicopter recovery force, requirements for communication, night vision capabilities, and joint interoperability.

RECOVERY FORCE

C-14. The PR force consists of the personnel affecting the actual recovery of the isolated personnel. This includes security personnel for the area around the extraction point; recovery personnel that authenticate and move the isolated personnel to the aircraft; and medical personnel that provide immediate assistance to the isolated personnel or injured security force personnel. The size and composition of this force may vary with the mission supported and the perceived or actual threat. During recovery operations, the RMC should be in the gun escort for SA at the objective; however, this is mission dependant.

Note. In medium aviation brigades, pathfinders generally conduct security force operations during personnel recoveries. When security forces from outside the CAB are utilized, it is even more critical to conduct training and rehearsals to ensure proper integration between participating units.

HELICOPTER RECOVERY FORCE

C-15. The helicopter recovery force consists of lift aircraft used to move the recovery force to and from the objective area and move the recovered IMDC personnel back to friendly forces. The helicopter recovery force will designate an AMC. The AMC coordinates all PR force efforts on the objective. The recovery force should include sufficient forces to move the isolated personnel to the recovery aircraft and provide immediate medical attention if required. If the situation dictates, a security force and personnel trained to remove injured personnel from wreckage may be added.

GUN ESCORT

C-16. The attack reconnaissance assets utilized to provide security escort to the helicopter recovery force may also be called the RESCORT. The primary duty of the gun escort or RESCORT is to provide protection and SA for the helicopter recovery force. The principles of air assault security are used in execution of this task. Priority is to avoid, suppress, and destroy targets posing a threat to the ground or helicopter recovery force, and to initiate communication with the OSC or IMDC personnel (if no OSC is on station).

PERSONNEL RECOVERY EXECUTION TASKS

C-17. The PR system is comprised of the five PR execution tasks. These tasks are central to any PR mission and must be accomplished to ensure a successful recovery.

REPORT

C-18. Timely, accurate reporting to the PR cell by subordinate units—

- Is the most critical and time sensitive PR action.
- Provides the commander options related to isolating incidents
- Suggests possible assets that may assist in recovery planning time.

C-19. Inaccuracies in reports such as call sign errors, location errors, unverified information, lack of personnel information, and lack of delegation within the staff causes problems that increase in magnitude due to the parallel nature of the PR planning process.

C-20. Historically, commanders have been hesitant to report a possible loss of accountability. Failure to report rapidly or have procedural controls in place to identify IMDC personnel causes a delay in responding. Rapid, coordinated action to recover the IMDC capitalizes on the enemy's relative confusion that generally surrounds an IMDC event.

LOCATE

C-21. If the IMDC location is unknown at the time of the initial report, every effort must be made to determine its location. Recovery efforts cannot commence until the IMDC's location is known. This highlights the criticality of gaining and maintaining SA. Once "eyes-on" and basic communication has been achieved, every effort must be made to maintain that level of SA. This requires an understanding of basic signaling and the management of assets to maintain contact.

SUPPORT

C-22. The support task includes actions taken to mentally, physically, and emotionally sustain IMDC personnel and their families, throughout the five tasks. Support to IMDC personnel includes establishing communications, conducting resupply, maintaining morale, and providing protection. While planning continues, IMDC personnel require support until units can conduct a recovery. IMDC personnel are authenticated to confirm their identities, family support is initiated with the rear detachment, and planning for the recovery is finalized and briefed to decision makers. All battlefield enablers should be employed to provide support to IMDC personnel. Creative use of "show CAS", FS, ISR, and CAB assets may be combined to ensure the isolated person understands he or she will be recovered.

RECOVER

C-23. The recover task includes the employment of forces to regain positive and procedural control of IMDC personnel and does not end until the IMDC personnel are handed over by the recovery element to medical personnel for reintegration. Once a COA has been approved, the order is passed to the recovery force. It may be necessary to launch the recovery force immediately to put them into a position to execute the recovery in a timely manner. The CAB commander generally retains the execute authority to ensure that more forces are not put at risk and all planning factors have been integrated to mitigate risk.

C-24. During the recovery, the PR cell continuously monitors the progress of the mission and keeps the PRCC informed to ensure that additional assets may be employed in a timely manner should the mission not progress as envisioned. Once the recovery of the IMDC has occurred, it is critical that medical/reintegration assets are updated and transportation is coordinated.

REINTEGRATE

C-25. PR execution does not stop with the successful recovery of IMDC personnel; it continues through the reintegration process. The goals of reintegration are two-fold—attend to the medical needs of the recovered personnel; and gather information about the event that has an immediate impact on current and future operations. The overriding concern during reintegration is the health and welfare of the recovered personnel. These factors take precedence over all others during the reintegration process. Reintegration team personnel must often balance these factors with the need to gather pertinent information from the recovered personnel. Reintegration also includes preparing IMDC for potential media interaction and providing other support to reduce their anxiety and possible frustration during recovery activities.

C-26. Every isolating incident affects individuals differently. Without medical and psychological assessments it is difficult to determine the exact affect. Commanders occasionally fear losing an individual to the process. This concern should be mitigated by the risk of sending personnel who have experienced emotional trauma back to duty.

Appendix D

Army Aviation Composite Risk Management

The tactical environment provides ever-changing demands and unpredictable problems, often under stressful conditions. The interface of man, machine, and environment is constantly shifting. We are challenged by a wide range of new technologies requiring our leaders to use creative measures to provide protection to our Soldiers on the battlefield. In this environment, mission accomplishment requires leader involvement and flexible decisionmaking. Not surprisingly, accidents and injuries increase during combat operations. Safety in the combat environment depends upon compliance with established standards; however, due to fluid conditions in the tactical environment, safe mission accomplishment relies heavily on complete integration of CRM into both planning and execution phases. CRM assists commanders in anticipating and controlling hazards in the planning phase and in dealing with unexpected hazards as they arise in the execution phase.

SECTION I – COMPOSITE RISK MANAGEMENT

GENERAL

D-1. CRM teaches Soldiers "how to think" rather than "what to think." Tough, realistic training conducted to standard is the cornerstone of Army warfighting skills. An intense training environment stresses both Soldiers and equipment, creating a high potential for accidents. The potential for accidents increases as training realism increases. Commanders must find ways to protect their Soldiers and equipment from accidents during realistic training to prepare for war. An accidental loss in war is no different in its effects from a combat loss; the asset is gone.

D-2. The CRC, located at Fort Rucker, Alabama, is the authority on CRM. Refer to FM 5-19 or visit the CRC website at https://crc.army.mil for more information.

THE PROCESS

D-3. The CRM process is defined in FM 5-19. CRM is the Army's primary decisionmaking process for identifying hazards and controlling risks across the entire spectrum of Army missions, functions, operations, and activities. CRM is a commonsense way of accomplishing the mission with the least risk possible. It is a method of getting the job done by identifying the areas that present the highest risk and taking action to eliminate, reduce, or control the risk. CRM thereby becomes a fully integrated part of mission planning and execution.

COMPOSITE RISK MANAGEMENT PLAN

D-4. CRM is an integral part of the MDMP. CRM is a decisionmaking process used to mitigate risks associated with all hazards that have the potential to injure or kill personnel, damage or destroy equipment, or otherwise impact mission effectiveness. The guiding principles of CRM are as follows:

- **Integrate CRM into all phases of missions and operations**. Effective CRM requires that the process is integrated into all phases of mission or operational planning, preparation, execution, and recovery.

- **Make risk decisions at the appropriate level**. As a decisionmaking tool, CRM is only effective when the information is passed to the appropriate level of command for decision. Commanders are required to establish and publish approval authority for decisionmaking. This may be a separate policy, specifically addressed in regulatory guidance, or addressed in the commander's training guidance. Approval authority for risk decisionmaking is usually based on guidance from higher headquarters.
- **Accept no unnecessary risk**. Accept no level of risk unless the potential gain or benefit outweighs the potential loss. CRM is a decisionmaking tool to assist the commander, leader, or individual in identifying, assessing, and controlling risks in order to make informed decisions that balance risk costs (losses) against mission benefits (potential gains).
- **Apply the process cyclically and continuously**. CRM is a continuous process applied across the entire spectrum of Army training and operations, individual and collective day-to-day activities and events, and base operations functions. It is a process that is used to identify and assess hazards, develop and implement controls, and evaluate outcomes.
- **Do not be risk averse**. Identify and control the hazards; complete the mission.
- **Knowledge of risk factors is a key to planning and decisionmaking**. With this knowledge leaders quantify risks, detect problem areas, reduce risk of injury or death, reduce property damage, and ensure compliance with regulations. Unit leaders should conduct risk assessments before conducting any training, operations, or logistics activities.

COMPOSITE RISK MANAGEMENT PLAN STEPS

D-5. CRM is a five-step process of assessment (steps 1 and 2) and management (steps 3 through 5) techniques used to eliminate or reduce hazards. The five-step process is—

- **Step 1** – Identify hazards.
- **Step 2** – Assess hazards to determine risk.
- **Step 3** – Develop controls and make risk decisions.
- **Step 4** – Implement controls.
- **Step 5** – Supervise and evaluate.

TRACKING AND DOCUMENTATION

D-6. To maintain continuity with mission tasks and requirements, it is necessary to track the CRM process in a standardized manner. Many tools are available that can be tailored to portray CRM information to suit a particular mission, situation, operation, or event. When time and situation allow, DA Form 7566 (Composite Risk Management Worksheet) or an electronic version is used to document the CRM process. In addition to providing an Army standard, continuous use of this worksheet reinforces CRM application and trains leaders, Soldiers, and individuals to think in terms of a five-step CRM process. Locally generated substitute forms, approved by individual commands, must contain as a minimum all information included on DA Form 7566.

SECTION II – OPERATIONAL TEMPO AND BATTLE RHYTHM

D-7. Battle rhythm allows units and leaders to function at a sustained level of efficiency for extended periods. Effective battle rhythm permits an acceptable level of leadership at all times. It can focus leadership at critical points in the fight or during particular events. Procedures and processes facilitating efficient decisionmaking and parallel planning are critical to achieving battle rhythm. Every component of battle rhythm contributes uniquely to sustained operations.

FATIGUE

D-8. Aviation operations are inherently dangerous. Commanders and leaders must be aware of increased dangers present in sustained operations. Additionally, leaders must be able to recognize the symptoms of chronic and acute fatigue and how to deal with fatigued aircrews. FM 3-04.301 provides an in-depth review of aero medical factors associated with aviation fatigue. Unit flight surgeons are trained to advise and assist the commander with managing or eliminating symptoms of fatigue.

FIGHTER MANAGEMENT

D-9. Fighter management describes a process essential to effective and efficient aircrew and battle staff operations. The aircrews are those personnel in the unit who execute the unit's flight mission. The battle staff are those personnel in the unit who manage the operations. Successful continuous operations at the unit level are sustained when commanders enforce rest periods. The unit requires an SOP defining rest periods, especially for crew members and mission critical personnel. The S-3 and operations personnel at battalion or higher levels can assist in monitoring, while commanders, flight surgeons and SOs manage this program. Commanders should utilize a tracking device to help manage a fighter management program. Table D-1 provides and example of the fighter management tracking system.

Table D-1. Example of fighter management tracking system

RANK:		*NAME:*			*SSN:*	
MONTH	Daily Totals		14-Day Totals		30-Day Totals	
	Reset w/10 hrs rest		Reset w/24 hrs rest		Reset w/48 hrs rest	
	Duty Hrs	Flight Hrs	Duty Hrs	Flight Hrs	Duty Hrs	Flight Hrs
Carry Fwd						
1						
2						
3						
4						
5						
6						
7						
8						
9						
10						
11						
12						
13						
14						
15						
16						
17						
18						
19						
20						
21						
22						
23						
24						
25						
26						
27						
28						
29						
30						
31						

D-10. AR 95-1 provides regulatory guidance for aviation duty periods and associated flight hours. Local commanders are the approval authority and designate duty hour, flight hour, and rest period requirements for their command. Table D-2, page D-4, provides an example of the duty period and flight-hour matrix.

Table D-2. Example of a duty period/flight-hour matrix

Time Period	*Duty Period*	*Flight Hours*
24 hrs	14 hrs	8-hour day (SR to SS) or *6 hrs-any combination (max 5 hrs may be N or NVD) 5 hrs-N or NVD
7 days	**84 hrs	42 Flight hrs
30 days		***90 Flight hrs
* Also includes Hood, weather, and MOPP. All MOPP level 3 or 4 flights will be factored at a rate of 2:1. MOPP training should not be conducted in wet bulb globe temperature above 85 degrees Fahrenheit.		
**If 84 hours are exceeded in a 7-day period, the individual will receive a 24-hour reset day before performing crew duties again.		
***Crew members reaching 90 hours within a 30-day period must consult with a flight surgeon before performing further flight duties. If a flight surgeon is not reasonably available, the crewmember may resume crew duties after receiving a minimum of 48 hours off-duty time by the commander. After resuming flight duties, the crewmember must be cleared by a flight surgeon every 10 hours as long as they exceed 90 flight hours in the current 30-day period.		

D-11. A 10-hour rest period is provided for aircrew members prior to beginning a new duty period. The rest period may be reduced to 8 hours for no more than 2 consecutive days by battalion/TF commander 0-5 or above). An 8-hour rest period is provided for nonflight personnel prior to beginning a new duty period.

D-12. If a 14-day duty or flight limit is reached, a 24-hour rest period is required. Any 24-hour rest period, within the 14th day duty or 48-hour rest period within the 30-day duty time, will reset the duty day time clock.

D-13. Extensions are not granted on a "blanket" basis; they are approved only on a "case-by-case" basis. Extension authority is outlined below:

- Company commander—can add 2 hours duty, not to exceed 14 hours, and one additional flight hour.
- Battalion/TF commander (0-5 or above—can add 2 hours duty not to exceed 16 hours and one additional flight hour.
- First 0-6 in the chain of command—can designate duty and flight hours as necessary (may only be deligated to 0-5 who is in command for a specified period).

SECTION III – AVIATION HAZARDS

D-14. Aviation operations involve inherently higher risk (higher probability of accidents and more severe consequences) than most ground operations. Historically, when deployed to combat theaters Army aviation has suffered more losses to accidents than to enemy action. Aviation accidents in combat are typically the same type experienced in peacetime. Because of this, commanders of units involved in aviation operations must emphasize the safety component of protecting the force. Commanders, supervisors, and safety managers at all levels must comply with certain policies regarding the aviation safety component for protecting the force as outlined in AR 385-10.

HAZARDS TO FLIGHT

D-15. Experience, judgment, maturity and discipline define today's Army aviators. Aviation commanders and leaders at all levels must continue to coach, teach and mentor one another. Aviation branch demands "standards". The following is a list of hazards and control measures to flight (table D-3, page D-5). This list is not all encompassing; however, by identifying hazards and controls associated with accidents that have already occurred, we can continue to learn from others and reduce overall aviation mishaps.

Table D-3. Hazards to flight

Hazards	***Controls***
Failure to estimate closure rates and necessary control inputs.	Maintain SA for proper distance estimation and closure rates.
Abrupt steep turns, high speed flight, low altitude and power management.	Maintain aircraft with in power limitations for conditions.
Inadequate pre-mission planning.	Plan and brief missions at the appropriate level.
Inadequate AMC duties.	Define and outline requirements for AMC program.
Zero illumination, combat formations, and unfamiliar areas.	Thorough planning, briefing and risk management.
Lack of mission training proficiency.	Conduct mission training to standard.
Lack of crew coordination.	Conduct aircrew coordination training enhanced (ACT-E).
Overconfidence and inadequate supervision.	Maintain unit discipline and leadership standards.
Engine failure.	Execute proper emergency procedure.
Inadequate scanning causes aircrews to impact wires, birds or flight hazards.	Conduct continuous scan and crew coordination.

Appendix E

Aircraft Characteristics

This appendix provides an overview of basic characteristics and capabilities of aircraft organic to aviation brigades or available in an aviation TF.

SECTION I – ROTARY WING AIRCRAFT

OH-58D (R) KIOWA WARRIOR

E-1. The primary missions of this aircraft are armed reconnaissance and attack. The OH-58D aircraft discussed herein are the version addressed with affectivity code "R" in TM 1-1520-248-10.

DESCRIPTION

E-2. The OH-58D is a single-engine, dual-seat, armed observation aircraft. It has an improved master controller processing unit system providing highly integrated communication, navigation, aircraft, and mission equipment subsystems. The video crosslink can store compressed images in memory and enables transmission of video images between aircraft. The mast-mounted sight (MMS) contains a suite of sensors including a high-resolution television camera, IR thermal imaging, laser rangefinder, laser designator, and videotape recorder. Table E-1 outlines OH-58D aircraft characteristics.

Table E-1. OH-58D characteristics

Specifications:
Length: 41 ft 2.4 in Height: 12 ft 10.6 in Fuselage width (w/weapons pylons): 9 ft 2 in Main Rotor Diameter: 35 ft Max Speed (Level): 110 kts Max Gross Wt: 5500 lbs (5200 lbs by interim statement of airworthiness qualification) Cruise Airspeed: 90 kts* Combat Radius: 120 km*
Armament:
Missile range (Hellfire): 8,000 m Missile range (ATA Stinger): 5,000 m Rocket range (Hydra 70): 6,000m (Airburst), 8000m (Contact)
Gun Range (.50 Cal):
2000 m (1600 m tracer burnout)
Optics - MMS:
TIS: Detection 10+ km Recognition 6-7 km Identification 3 km
Television Sensor:
Detection 8+km Recognition 7 km Identification 4-6 km

Table E-1. OH-58D characteristics

Laser Range Finder/Designator:
Maximum ranging distance 9.99 km. Lasing a known point will update the navigation system. Maximum designating distance limited only by TIS/television sensor.
Navigation Equipment:
Inertial Navigation System (INS)/GPS. Can slave MMS to grid input by operator.
* varies with environmental/mission conditions

CAPABILITIES

E-3. The OH-58D provides the following:

- Day, night, battlefield obscurant, and limited adverse-weather fighting capabilities.
- Data transfer system permitting upload from AMPS DTM and downloading of selected postmission data.
- Countermeasure suite of IR jammers, radar warning receivers, and laser warning detectors.
- Moving map display.
- Video recording and cockpit playback of television and thermal imagery from the mission.
- Advanced navigation and mission planning equipment; transportable in the C-130, C-141, C-5, and C-17.

ARMAMENT SYSTEMS

E-4. The OH-58D armament capabilities consist of a .50-caliber machine gun, 2.75-inch rockets, Hellfire missiles, and Stinger air-to-air missiles. These systems are mounted on two universal weapons pylons. The aircraft has a laser rangefinder/designator used to designate for the weapons system as well as provide range-to-target information for onboard weapons systems. Additionally, the OH-58D utilizes an AIM-1 IR Laser Aiming Light as an aiming device for night combat engagements. The AIM-1 can also be utilized to identify and/or confirm enemy targets.

.50-Caliber Machine Gun

E-5. The machine gun is an air-cooled, belt fed, recoil operated, electronically controlled weapon. The gun is mounted in a fixed position to the universal weapon pylon on the left side of the aircraft. It is capable of firing 750-850 rounds per minute at a maximum effective range of 2,000m. The ammunition feed and storage system holds approximately 500 rounds.

2.75-Inch Rocket System

E-6. The 2.75-inch weapon system is a light assault weapon for use against enemy personnel, light armored vehicles, and other soft-skinned targets. The system is comprised of one or two 7-shot rocket launchers and may be installed on one or both sides of the aircraft. This area system can launch multiple rockets with various warhead mixes including high explosive, high-explosive MPSM, white phosphorous, illumination, and flechette. The maximum range is 6,000m for airburst warheads, and 8,000m for contact warheads.

Hellfire Missile

E-7. The Hellfire missile system is an air-to-surface, laser guided missile system. The hellfire missile is a point target weapon system using high explosive anti-tank (HEAT) warheads to destroy armored targets and blast fragmentation and thermo-baric warheads to defeat all targets, except armor targets, utilizing blast and fragmentation. The Hellfire missile system can be comprised of one or two launchers containing up to two missiles each. However, weight restrictions may limit the aircraft to just one launcher. The minimum engagement range is 500 meters; maximum range is 8,000 meters. Laser designation may be autonomous or by remote ground or airborne designators.

Air-to-Air Stinger Missile System

E-8. The air-to-air Stinger is an IR, heat-seeking, fire-and-forget missile capable of engaging airborne targets day or night. The OH-58D can carry two Stinger missiles per pylon for a maximum of four missiles. The maximum range is more than 4,000 meters.

ARMAMENT CONFIGURATIONS

E-9. Table E-2 provides typical OH-58D ordnance loads.

Table E-2. Typical OH-58D helicopter ordnance loads

Aircraft	*Gun*	*2.75-in Rockets*	*Missiles*
OH-58D	500	7	
OH-58D		14	
OH-58D	500		2 air-to-air Stingers
OH-58D			4 Hellfire

E-10. Figure E-1 illustrates an example of mission load configurations. Only one system at a time may be mounted per side.

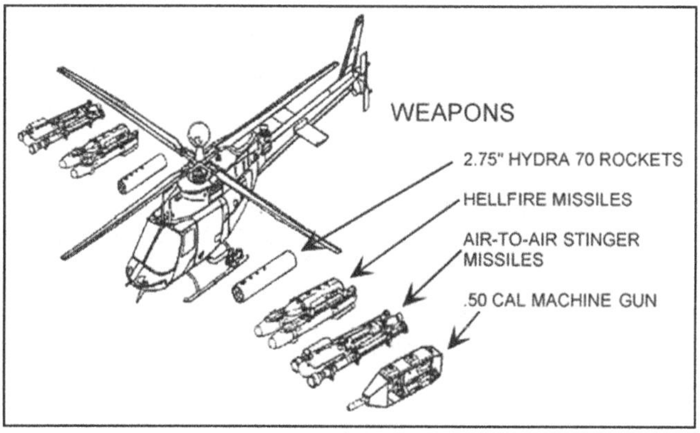

Figure E-1. OH-58D weapons loading

COMMUNICATIONS

E-11. The OH-58D has the following communications systems:

- Two AN/ARC-201D SINCGARS with embedded encryption and data capability. Both operate in the 30 to 87.975 MHz frequency range and have antijam, frequency-hopping capability.
- The AN/ARC-186 provides VHF-AM (116 to 151.975 MHz).
- The AN/ARC-164 Have Quick II provides antijam frequency-hopping UHF-AM communications in the 225 to 399.975 MHz range at 25 KHz intervals.
- An IDM transfers digital messages. The KY-58 provides secure communications for Have Quick II and VHF-AM/FM radios.

NAVIGATION SYSTEM

E-12. The embedded global positioning system/inertial navigation system (EGI) is a self-contained, all-attitude navigation system that works with the radar altimeter and laser rangefinder/designator. Its embedded GPS receiver supports both pure and blended GPS, and INS solutions. The system is capable of storing waypoint and target information for making flight plan routes and altering these routes if a mission changes. EGI also provides target location in longitude and latitude or MGRS coordinates and altitude in meters or feet MSL when range, bearing, and declination to the target are provided to the navigation subsystem for the MMS.

LIMITATIONS

E-13. The following paragraphs discuss limitations of the OH-58D.

Infrared Radiation Crossover

E-14. The thermal imaging sensor operates by determining temperature differentials; when targets and their surroundings reach the same temperature, target detection degrades.

Obscurants

E-15. Some obscurants (dust, rain, haze, or smoke) can keep laser energy from reaching the target and hide it from the incoming munitions seeker.

Low Cloud Ceilings

E-16. Consideration must be given to cloud ceilings determining maximum employment range of the SAL Hellfire. Depending on the distance to target, trajectory mode selected, and lasing techniques, the missile may climb into low cloud ceilings causing the seeker to break track from the laser spot or preventing the seeker from acquiring the laser spot.

Hellfire Remote Designation Constraints

E-17. The designating crew may offset a maximum of 60 degrees from the gun-to-target line and must not position its aircraft within a 30-degree safety fan from the firing aircraft.

Instrument Meteorological Conditions

E-18. The OH-58 Kiowa Warrior is not certified for IFR operations.

APACHE (ALPHA/DELTA)

E-19. The AH-64 is a highly stable aerial weapons-delivery platform. Its primary role is to attack. It can fight to destroy, attrit, disrupt, or delay enemy forces. Armed reconnaissance and security are secondary roles.

DESCRIPTION

E-20. The AH-64 is a twin-engine, tandem-seat, aerial weapons platform. It is equipped with a target acquisition designation sight (TADS), tri-service laser designator/range finder, and a pilot night vision sensor (PNVS) allowing the two-man crew to navigate and attack at night at extended standoff ranges, and in some adverse weather conditions. The Apache has a full range of ASE.

E-21. The AH-64D is a remanufactured AH-64A. Its improved navigation system integrates dual INS/GPS and Doppler radar for acceleration cueing. Some Longbow models are equipped with a millimeter wave fire control radar (FCR) allowing the helicopter to identify, classify, prioritize, and track certain targets to the maximum range of the Hellfire weapon system. Table E-3 compares Apache aircraft specifications. Table E-4 details AH-64D characteristics.

Table E-3. Comparison of Apache specifications

Model	***AH-64A***	***AH-64D w/o Radar***	***AH-64D Longbow***
Length (ft)	57.67	57.67	57.67
Height (ft)	15.25	13.33	16.08
Width (ft)	17.17	15.50	15.05
Main Rotor Span (ft)	48	48	48
Max Gross Wt (lbs)	21,000	23,000	23,000
Cruise Speed (kts)	120*	130*	130*
Combat Radius (km)	200*	200	200
Combat Radius w/One 230-Gallon Aux Fuel Tank (km)	350*	350*	350*
Self-Deployability	Yes	Yes	Yes
*Varies with a multitude of factors such as temperature, wind, gross weight, and mission-specific time requirements.			

Table E-4. AH-64D characteristics

Armament:
Maximum autonomous Hellfire missile range: 7,000m Maximum remote designated Hellfire missile range: 8,000m 2.75" Rocket hover fire most effective dispersion pattern: 3,000 – 5,000m for MPSM and unitary warhead 2.75" Rocket running/diving fires in CCA accuracy degrades beyond 1,000-1,200m Maximum 30mm Gun range: 4,000 m Maximum effective gun range from a hover using TADS: 1,500 point targets, 3,000m area targets Running/diving fires in CCA accuracy degrades beyond 800-1,000m
Optics:
TADS/Day TV (Low Light, Daytime) Allows autonomous laser designation of tank size target with image auto tracker at 6,000m, 4,000m using manual tracking TADS/TIS (day, night, weather, obscurants) Allows autonomous laser designation of tank sized target with image auto tracker at 3,500m, 3,000 using manual tracking. PNVS (night, weather, obscurants) MTADS (Second generation TIS) greatly increases TADS capability to detect and identify, not available on all AH-64D aircraft.
Navigation Equipment:
Dual EGI/GPS, Doppler radar, ADF
Flight Characteristics:
Max speed in level flight is a function of environmental conditions, aircraft weight and engine capabilities and will not normally exceed 140-145 Kts.. Normal cruise speed: 110 - 120 kts
Additional Capabilities:
Aircraft can be configured with up to four 230 gallon external fuel tanks for ferrying/self-deployment missions. Use of IZLID laser for AGI to allow ground units to view laser through NVDs. Can be configured with Roberson internal fuel tank to increase flight endurance by 45-55 minutes.
Limitations:
Threat ID; IR crossover; Weather may inhibit Hellfire engagements (seeker must be able to "see" the laser designated spot); overwater operations severely degrade navigation system; PNVS cannot detect wires or other small obstacles. TADS and PNVS TIS quality can be severely degraded by weather conditions that will not normally affect ground maneuver forces, such as rain, fog, snow. Aircrews may or may not have NVG capability.

Fire-Control Radar

E-22. The Apache Longbow system consists of an integrated millimeter wave FCR, along with a radio frequency interferometer (RFI). The FCR enables Apache Longbow helicopters to detect, classify, prioritize, and engage targets with radar frequency Hellfire missiles without visually acquiring the target. Apache Longbow crews may also employ radar frequency Hellfire missile during poor visibility when laser, optical, and TIS sensors are degraded. The FCR will not identify friend or foe, however the RFI can detect and identify radar systems and display targeting information on the same screen as the information from the FCR.

E-23. Apaches equipped with the Longbow system are denoted as either AH-64D with radar or AH-64D Longbow. The Longbow is equipped with FCR, a radar frequency interferometer, and upgraded 701C engines to compensate for the additional weight of the Longbow system. The Longbow system is integrated with the TADS to allow simultaneous and autonomous operation of the TADS and the FCR.

Without Radar

E-24. The AH-64D without radar includes all preceding Longbow aircraft upgrades except an FCR, a radar frequency interferometer system, and their associated black boxes. The aircraft may not have the improved 701C engine installed. The AH-64D without radar can be converted to an AH-64D with radar upon installation of the Longbow system and 701C engines. The AH-64D without radar can fire the radar frequency Hellfire missile autonomously (with LOS to the target) or by using FCR targeting data handed over from an AH-64D with radar.

CAPABILITIES

E-25. The AH-64 provides the following:

- Precision attacks during day or night, or when the battlefield is obscured (Longbow).
- Wide array of firepower options.
- Detection, classification, and prioritization of stationary and moving ground and airborne targets (Longbow).
- Robust suite of EW systems.
- Lethal destruction of enemy ADs.
- Real-time SA and intelligence of the battlefield to the digitized aviation/ground commander; data transfer system to upload from the AMPS data-transfer cartridge and download postmission data.
- HF radio for NOE long-distance NLOS communications.

ATTACK HELICOPTER ARMAMENT SYSTEMS

E-26. The AH-64 can carry up to a total of 16 Hellfire laser-designated missiles on four wing store pylons, or up to a total of 76 2.75-inch folding fin aerial rockets on four wing store pylons, and up to 1,200 rounds of ammunition for its 30-millimeter cannon. Wing stores may be configured to allow both hellfire and rockets to be carried. Each pylon may carry up to 19 rockets or 8 hellfire missiles. Environmental conditions may significantly restrict weapons loads and station time associated with those loads. In addition, the use of a Roberson internal auxiliary fuel tank will limit 30mm to 300 rounds.

E-27. The Apache Longbow system enhances the rapid employment of all available weapons including Hellfire missiles, aerial rocket system, and the 30-millimeter cannon. Once the FCR detects, classifies, and prioritizes targets, the gunner selects the desired weapon for attack. These data are automatically transferred to the weapon and displayed on the selected weapon sight.

30-Millimeter Chain Gun

E-28. The M230E1 is a chain-driven area weapons system mounted to a hydraulically driven turret under the helicopter forward fuselage. It fires the U.S. M789 NATO standard ammunition HEDP round. It can penetrate more than 2 inches of armor at 2,500 meters and produces antipersonnel effects within a 4-meter radius. At typical engagement ranges, HEDP ammunition will defeat BMP type targets. The AH-64 can carry 1,200 rounds of 30-millimeter ammunition. It has a rate of fire of 600 to 650 rounds per minute with a maximum effective range from a hover using TADS of approximately 1,500 meters against point targets and 3,000 meters against area targets. Running/diving fires in a CCA environment are limited by range source selection and often fired using the helmet sight at ranges of less than 1,000m.

2.75-Inch Rocket System

E-29. When configured with four 19 shot rocket pods, the AH-64D can carry a maximum of 76 folding-fin aerial rockets for use against enemy personnel, light armored vehicles, and other soft-skinned targets. The system can launch multiple rockets with various warhead mixes to include; high explosive point detonating, high-explosive MPSMs, white phosphorous, red phosphorous, overt and covert illumination, and flechette. Aircrews select the quantity and type to be fired. The maximum range varies with warhead. If configured with four rocket pods, hellfire launchers are not installed. Environmental conditions may significantly restrict weapons loads and station time.

Hellfire Missile

E-30. The Hellfire is used primarily to destroy tanks, armored vehicles, and other hard-material targets. The AH-64A is capable of firing only the SAL Hellfire while the AH64D is capable of firing both the SAL and radar frequency Hellfire missiles. SAL and radar frequency Hellfire can defeat any known armor. The minimum engagement range is 500 meters, the maximum range is 8,000 meters, and the maximum aircraft load when configured with four hellfire launchers is 16 missiles. If configured with four hellfire launchers, 2.75-inch rocket pods are not installed. Environmental conditions may significantly restrict weapons loads and station time.

Semiactive Laser Hellfire

E-31. The SAL Hellfire requires a laser target designation that may be autonomous or remote

E-32. The remote designator may offset a maximum of 60 degrees from the gun-to-target line and must not position its self within a 30 degree safety fan from the firing aircraft.

Longbow Radar Frequency Hellfire

E-33. The Longbow radar frequency Hellfire is a millimeter wave guided missile and a true fire-and-forget weapon. The millimeter wave radar and missile can engage targets through weather and battlefield obscurants. The radar frequency missile receives targeting information (to include north, east, and down data) from the acquisition source—TADS, FCR, or another aircraft. Targeting data can be transferred from a Longbow to an AH-64D without radar as a radar frequency handover.

MISSION CONFIGURATIONS

E-34. Table E-5 is a matrix of AH-64D mission profiles and typical ammunition loads (weight limits may require reduction in mission loads).

Table E-5. AH-64D weapons loads, weights, and radius

Weapons:	*A*	*B*	*C*		*Weights & Radius: (110 kts airspeed)*	*A*	*B*	*C*
Hellfire	8	16	12		Operate	13897	14009	13953
Rockets	38	0	19		Fuel	2870	2870	2870
30mm	340	340	340		Load	2103	1862	1982
					T/O	18870	18741	18805
					Radius KM	210.6	210.6	210.6
					Radius NM	113.7	113.7	113.7
					Endurance	2.23	2.23	2.23

COMMUNICATIONS

E-35. The AH-64D has the following communications systems:

- The AN/ARC-201D SINCGARS with embedded encryption and data capability (operates in the 30 to 87.975 MHz frequency range and has antijam, frequency-hopping capability).
- The AN/ARC-164 Have Quick II provides antijam frequency-hopping UHF-AM communications in the 225 to 399.975 MHz range at 25 KHz intervals.
- The AN/ARC-186 provides VHF-AM (116 to 151.975 MHz).
- The AN/ARC-220 HF radio provides NOE long-range communications in the 2 to 29.999 MHz range and secure mode when employed with the KY-100.
- The TSEC/KY-100 provides secure communications for the AN/ARC-220 HF radio.
- The TSEC/KY-58 interfaces with AN/ARC-201D and AN/ARC-201 radios to provide secure voice for these radios.
- The IDM transfers digital messages.
- Some aircraft may have a second SINCGARS radio in place of the HF radio.

NAVIGATION SYSTEMS

E-36. The navigation subsystem consists of the following major components:

- Embedded global positioning system/inertial navigation system (EGI), primary and backup.
- Doppler radar velocity sensor.
- Radar altimeter.
- Automatic direction finder (ADF).
- High integrated air data computer.
- Flight management computer.

LIMITATIONS

E-37. Prominent limitations of the AH-64 are discussed in the following paragraphs.

Threat Identification

E-38. Threat identification through the TIS is extremely difficult. Although the crew can easily find the heat signature of a vehicle, it may not be able to determine whether it is friend or foe. FCR target identification is limited to radar cross section return data and does not determine actual target validity.

Infrared Radiation Crossover

E-39. The TIS and PNVS operate by determining temperature differentials. When targets and their surroundings reach the same temperature (normally twice a day), target detection is degraded; these conditions make flight difficult while using the TIS. The same effect occurs when temperatures do not significantly vary throughout the day or night, or heavy cloud cover exists for long periods of time. For example, sustained rainfall may reduce the TIS quality to unusable for flight or targeting during certain times of the year.

Obscurants

E-40. Some obscurants (such as dust, rain, haze, or smoke) can prevent laser energy from reaching the target. It also hides the target from incoming munitions seekers for SAL Hellfire and prevents effective use of TIS. AH-64D FCR and radar frequency Hellfire see and shoot through obscurants.

Low Cloud Ceilings

E-41. Determination of the SAL Hellfire's maximum employment range requires consideration of cloud ceilings. Depending on range to target, trajectory mode selected, and lasing techniques, the missile may climb into low cloud ceilings causing the seeker to break track from the laser spot or preventing the seeker from acquiring the laser spot.

Instrument Flight Rules

E-42. The AH-64D is not currently certified for IFR operations.

BLACK HAWK

E-43. The primary missions of this aircraft are air assault, air movement, and C2 Support. Additional roles include PR, aircraft recovery, parachute operations, disaster relief, and fire fighting.

DESCRIPTION

E-44. The UH-60A/L is a twin-engine, dual-seat, utility helicopter. The minimum required crew is a pilot and copilot. It is designed to carry 11 combat-loaded air assault troops (seats installed). It also can move a 105-millimeter howitzer and 30 rounds of ammunition. The UH-60A/L is equipped with a full instrument package and certified for IMC as well as day and NVG operations. Table E-6 provides UH-60A/L aircraft characteristics.

Table E-6. UH-60A/L aircraft characteristics

Specifications:	
Length: 64 ft 10 in rotors turning, 41 ft 4 in rotors/pylons folded. Height: 12 ft 4 in center hub, 16 ft 10 in tail rotor. Width: 9 ft 8.6 in main landing gear, 14 ft 4 in stabilator. Width with external stores support system (ESSS) installed: 21 ft. Main rotor and tail rotor diameter: 53 ft 8 in main rotor, 11-ft tail rotor at 20-degree angle. Cabin floor dimensions: 73 in wide x 151 in long.	Cabin door dimensions: 69 in wide x 54.5 in high. Maxi gross weight, UH-60A/L: 22,000 lbs*. Maximum cargo hook load, UH-60A: 8,000 lbs*. Maximum cargo hook load, UH-60L: 9,000 lbs*. Cruise airspeed: 130 kts *. Combat radius: 225 km *.
Armament:	
2 x M60D or M240 (7.62 MGs) (self protection only).	
Optics:	
Pilots use AN/AVS-6 to fly the aircraft at night.	
Navigation Equipment:	
Doppler/GPS navigation set.	
Flight Characteristics:	
Max speed (Level): 156 kts. Normal cruise speed: 120-145 kts. With external slingloads: 140 kts max. Up to 8,000 lbs/120 kts max. 8,000-9,000 lbs.	
Additional Capabilities:	
The ESSS allows configuration for extended operations without refueling (5+ hours) (2 X 200 gallon fuel tanks) and ferry and self-deployment flights (4 X 200 gallon fuel tanks). The enhanced C2 console provides the maneuver commander with an airborne platform supporting six secure FM radios, one HF radio, two VHF radios, and two UHF radios. Can be configured with the Volcano Mine Dispensing System; requires 8 hours to install. Capable of inserting and extracting troops with FRIES/SPIES.	
Limitations:	
Use of the ESSS for fuel limits access to the cabin doors for troops and bulky cargo or litters. It also greatly decreases the payload and maximum speed. UH-60A cannot sling-load a TOW HMMWV. Cruise speed is greatly decreased by light, bulky sling-loads (less than 80 kts).	
* varies with environmental/mission conditions	

CAPABILITIES

E-45. The UH-60A/L provides the following:

- Countermeasure suite of IR jammers and radar warning receivers.
- Data-transfer system to upload from the AMPS data-transfer cartridge and download postmission data.
- Internal transport of 11 combat-loaded troops with seats installed and approximately 16 combat-loaded troops with seats removed. Actual number of troops carried is limited by space and environmental conditions.
- Self-deployable range of 558 NM with the ERFS, with 30-minute reserve.
- Transportable by the C-5 and C-17 aircraft.

ARMAMENT SUBSYSTEMS

E-46. The UH-60 has provisions for door mounting of two M60D/M240 7.62-millimeter machine guns. The subsystem is pintle-mounted in each gunner's window at the forward end of the cabin section. The two machine guns are free pointing but limited in traverse, elevation, and depression.

AIR VOLCANO

E-47. The air Volcano is a helicopter-mounted, automated, scatterable mine-delivery system able to deliver mines day or night. The system can rapidly emplace a 278-meter, 557-meter, or 1,115- by 140-meter field at up to 960 mines (800 AT and 160 anti-personnel) per sortie. AT density yields an 80 percent chance of encounter. Mines can be set to self destruct after 4 hours, 48 hours, or 15 days. The air Volcano system has the following limitations:

- The UH-60 with air Volcano mounted, a full crew, and one system operator will be at high gross weight, which reduces range and maneuverability.
- Minefield emplacement is conducted at low airspeeds (80 kts or less), making the aircraft more vulnerable to detection and engagement.
- Crew cannot operate the M60D/M240 machine gun with the air Volcano installed.
- Installation requires approximately 8 hours.
- Requires two 5-ton cargo trucks for transport; it is an engineer responsibility to provide transportation assets to move these systems.

E-48. Four types of minefields can by emplaced using Volcano—disrupt, fix, turn, and block. Refer to FM 1-113 for additional information.

COMMUNICATIONS

E-49. The UH-60 A/L has the following communications systems:

- The AN/ARC-186 provides two-way voice communications in both the VHF-AM-FM ranges. It provides VHF-AM ATS communications.
- The AN/ARC-164 (V) Have Quick II provides two-way voice communications in the UHF-AM frequency range of 225 to 399.975 MHz; the sets provide an antijam frequency-hopping capability.
- The AN/ARC-201 (SINCGARS) is a VHF-FM antijam frequency-hopping radio, providing communications in the 30 to 87.975 MHz frequency range at 25 KHz intervals.

- The AN/ARC-220 HF radio provides NOE, long-range communication with the AN/ARC-100 in the CP and with other HF receivers.
- The TSEC/KY-58 interfaces with the ARC-186 (V), Have Quick II, and SINCGARS radios to provide secure communications.

NAVIGATION SYSTEMS

E-50. The UH-60A/L has the following navigation systems:

- The ASN-128B/D Doppler/GPS navigation set provides present position or destination navigation information in latitude and longitude or MGRS coordinates.
- The AN/ARN-89 or AN/ARN-149 (V) provides automatic direction-finding capability for instrument navigation and approach.
- The AN/ARN-123 (V) or AN/ARN-147 (V) VOR/LOC/GS/MB receiving sets provide instrument navigation and approach.

HEADS-UP DISPLAY AN/AVS-7

E-51. The heads-up display (HUD) system serves as an aid to pilots using the AN/AVS-6 NVG by providing operational symbology information directly into the NVG. It always displays airspeed, altitude from MSL, attitude, and engine torque and can display up to 29 symbols.

LIMITATIONS

E-52. The following are limitations of the UH-60A/L aircraft:

- UH-60A/L aircrews employ AN/AVS-6 NVG that lack the same night capabilities as AH-64 and OH-58D TISs.
- UH-60A/L aircraft are instrument-certified but cannot operate in all environmental conditions.
- Aircraft equipped with extended-range fuel tanks may not offer the same accessibility to the aircraft cabin for loading; self-defense machine guns have a limited range of motion when ERFS kits are installed.

HH-60L BLACK HAWK

E-53. The HH-60L Black Hawk's primary mission is aeromedical evacuation. Secondary missions include transport of medical personnel and equipment, emergency transport of class VIII to include blood products and biologicals, PR support, and support to Title 10 U.S. Code taskings.

DESCRIPTION

E-54. The HH-60L is a twin-engine, dual seat, utility helicopter. The minimum required crew is a pilot and copilot. For aeromedical evacuation missions, the crew includes up to three medical attendants (typical configuration includes one crew chief and two medical attendants). The HH-60L is equipped with a full instrument package and certified for IMC, as well as day and NVG operations. In addition to its basic configuration, the HH-60L includes a nose mounted TIS and kit installations allowing rescue hoist, extended-range fuel, and aeromedical evacuation operations. The interior design of the HH-60L allows for the installation of life-saving instruments and equipment for use by on-board medical attendants. Normal cabin configurations of the HH-60L can accommodate up to four primary litter patients or six ambulatory (seated) patients. However, when necessary, two Standardization Agreement (STANAG) litters can be placed on the floor under the forward lifts for a total of six litter patients. The unique platform design also includes oxygen distribution and suction systems, an airway management capability, and provisions for stowing intravenous solutions. The interior also features the following capabilities essential to providing the highest degree of patient care when every second counts:

- Oxygen generating systems.
- NVG compatible lighting throughout.

- An environmental control system.
- Medical equipment.
- Patient monitoring equipment.
- Neonatal isolettes.

E-55. Table E-7 outlines HH-60L aircraft specifications.

Table E-7. HH-60L specifications

Length	64 ft 10 in rotors turning, 41 ft 4 in rotors/pylon folded
Height	12 ft 4 in center hub, 16 ft 10 in tail rotor
Width	9 ft 8.6 in main landing gear, 14 ft 4 in stabilator
Main rotor & tail rotor diameter	53 ft 8 in main rotor, 11 ft tail rotor at 20-degree angle
Cabin floor & door dimensions	73 in wide x 151 in long, 69 in wide x 54.5 in high
Maximum gross wt	22,000 lbs
Rescue hoist/cargo hook max wts	600 lbs rescue hoist; 8,000 lbs, cargo hook
Maximum range w/ERFS	630 NM w/400 lbs reserve
Patient capacity	6 litter or 6 ambulatory
Crew capacity	2 pilots, 1 crew chief, 2-3 medical attendants
Fuel capacity	360 gallons and additional 400 gallons w/ERFS

CAPABILITIES

E-56. The HH-60L provides transport of the following:

- Six litter patients and two medical attendants.
- Six ambulatory patients and two medical attendants.
- Internally and externally loaded medical supplies.
- Medical personnel.

ARMAMENT

E-57. The HH-60L is an unarmed aircraft (no gunner windows).

COMMUNICATIONS

E-58. The HH-60L has the following communication systems:

- The AN/ARC-201 (SINCGARS) provides VHF-FM communications in the 30 to 87.975 MHz frequency range and has antijam, frequency-hopping capability.
- The AN/ARC-222 provides VHF-AM/FM communications and a maritime capability.
- The ARC-164 (V) provides UHF-AM communications in the 225 to 399.975 MHz frequency range and has antijam, frequency-hopping capability.
- The AN/ARC-220 HF radio provides NOE long-range communications in the 2 to 29.999 MHz frequency range.
- The KY-58 provides secure communications for the SINCGARS, Have Quick II, and AN/ARC-222 radios.
- The KY-100 provides secure communications for the AN/ARC-220 high frequency radio.

NAVIGATION SYSTEMS

E-59. The HH-60L has the following navigation systems:

- AN/ARN-149 ADF.
- AN-ARN-147 VOR/LOC/GS/MB receiving set.

- AN/ASN 128D Doppler/GPS navigation set.
- AN/ASN 153 (V) tactical air navigation.

AN/ARS-6(V) Pilot Heads-up Display AN/AVS-7

E-60. The HUD system serves as an aid to pilots using the AN/AVS-6 NVG by providing operational symbology information directly into the NVG. It always displays airspeed, altitude (MSL), attitude, and engine torque and can display up to 29 symbols.

Limitations

E-61. The following are limitations of the HH-60L aircraft:

- HH-60L aircrews employ AN/AVS-6 NVG that lack the same night capabilities as AH-64 and OH-58D TISs. The HH-60L TIS is for mission detection of Soldiers to be evacuated. It is not compatible for flying the aircraft.
- HH-60L aircraft are instrument certified but cannot operate in all environmental conditions, depending on threat and NAVAID availability.
- Aircraft equipped with extended-range fuel tanks may not offer the same accessibility to the aircraft cabin for loading.
- The medical equipment in the HH-60L is hard-mounted and cannot be removed for conversion to a UH-60L.

CHINOOK

E-62. The primary missions of this aircraft are air assault and air movement. Additionial roles include PR, CASEVAC, aircraft recovery, parachute operations, disaster relief, and fire fighting.

Description

E-63. The CH-47D is a twin-turbine engine, tandem-rotor helicopter. The minimum crew required to fly it is a pilot, copilot, and flight engineer. Additional crew members, as required, may be added at the discretion of the commander. Tactical missions normally require the addition of one or two crew chiefs. Table E-8 outlines CH-47D characteristics.

Table E-8. CH-47D characteristics

Specifications:	
Length: 98.9 ft Height: 18.9 ft Fuselage width: 12.4 ft Main rotor span: 60 ft Cargo space: 1,500 cu ft Floor space: 225 sq ft Maximum gross weight:50,000 lbs	Max load for forward and aft hooks: 17,000 lbs Max tandem load for forward and aft hooks: 25,000 lbs Max load for center hook: 26,000 lbs Cruise airspeed: 130* kts Max continuous airspeed: 170* kts Combat radius (16,000 lbs cargo): 50* NM (90 km) Combat radius (31 troops): 100* NM (180 km)
Armament:	
2 M60D 7.62 MGs (self protection only).	
Optics:	
Pilots use AN/AVS-6 to fly the aircraft at night.	
Navigation Equipment:	
Doppler/GPS navigation set.	
Flight Characteristics:	
Max Speed (Level): 170 kts Normal Cruise Speed: 120-145 kts	

Table E-8. CH-47D characteristics

Additional Capabilities:
Can be configured w/additional fuel for mobile FARE system (Fat Cow) or for ferrying/self-deployment missions. Aircraft has an internal load winch to ease loading of properly configured cargo. The CH-47D can sling-load virtually any piece of equipment in the Light Infantry, Airborne or Air Assault Divisions.
Limitations:
Cruise speed is greatly decreased by light, bulky sling-loads (less than 80 kts).
*Varies with factors such as temperature, wind, gross weight, internal versus external load, and time in PZ/LZ.

Capabilities

E-64. The CH-47D provides the following:

- Countermeasure suite of IR jammers, radar-warning receivers, and laser-warning detectors.
- Data-reduction transfer system to upload from the AMPS data-transfer cartridge and download postmission data.
- Internal transport of two HMMWVs or a HMMWV with a 105-millimeter howitzer and gun crew.
- Evacuation of 24-litter patients and 2 medics.
- Self-deployable range of 1,056 NM with the ERFS, with 30-minute reserve; transportable in the C-5 aircraft.

Armament Subsystems

E-65. The armament subsystems are M24 and M41 machine-gun systems installed in the cabin door, cabin escape hatch, and on the ramp. Both subsystems use the M60D 7.62-millimeter machine gun. The two flexible 7.62-millimeter machine guns are free pointing but limited in traverse, elevation, and depression.

Communications

E-66. The CH-47D has the following communications systems:

- The AN/ARC-164 Have Quick II radios provide UHF-AM two-way communications in the 225 to 399.975 MHz range in 25 KHz intervals; they can operate in normal or antijam, frequency-hopping mode.
- The AN/ARC-201 SINCGARS provides two-way communications in the VHF-FM range of 30 to 87.975 MHz in 25 KHz intervals; it employs antijam, frequency-hopping capability, and, when used with the KY-58, provides secure voice and cipher-mode communications. Later SINCGARS has embedded encryption and does not require KY-58 interface.
- One or two AN/ARC-186 VHF-AM-FM radio sets are installed, providing broad VHF communications on either the number 1 or 3 position on the function control selector of the controls and function, interphone control.
- The AN/ARC-220 HF radio supports NOE long-distance communications from 2 to 29.999 MHz in 100-hertz steps on 20 preselectable channels, for a total of 280,000 possible frequencies; as one of the radios available to the commander, it is accessible in the number 4 position on the function control selector.
- The KY-58 interfaces with the AN/ARC-186 VHF-AM-FM radio in the FM range to provide secure communications.
- The KY-100 provides secure communications for the AN/ARC-220 high frequency radio.

NAVIGATION SYSTEMS

E-67. The CH-47 has the following navigation systems:

- The AN/ASN-128B Doppler/GPS navigation set provides present position or destination navigation information in latitude and longitude or the MGRS coordinates. In the primary combined mode, the GPS updates Doppler position at a 1-MHz rate; other CH-47 aircraft have the AN/ASN-128 Doppler navigation without GPS.
- The AN/ARN-89 ADF provides automatic direction finding for instrument navigation and approach.
- The AN/ARN-123 (V) VOR/LOC/GS/MB provides instrument navigation and approach.

HEADS-UP DISPLAY AN/AVS-7

E-68. The HUD system serves as an aid to pilots using the AN/AVS-6 NVG by providing operational symbology information directly into the NVG. It always displays airspeed, altitude (MSL), attitude, and engine torque and can display up to 29 symbols.

AIRCRAFT SURVIVABILITY EQUIPMENT

E-69. Refer to FM 1-113 for detailed information on CH-47D ASE.

LIMITATIONS

E-70. The following are limitations of the CH-47D aircraft:

- CH-47D aircrews employ AN/AVS-6 NVG that lack the same night capabilities as AH-64 and OH-58D TISs.
- CH-47D aircraft are instrument certified but cannot operate in all environmental conditions.

TYPICAL FUEL EXPENDITURE RATES, CAPACITIES, AND STANDARD LOAD CAPACITIES

E-71. Table E-9 depicts typical rates of fuel expenditures per helicopter and fuel capacities without additional tanks.

Table E-9. Typical helicopter fuel expenditure rates and capacities

Helicopter	*Average Gallons per Hour*	*Fuel Capacity*
AH-64D	175	370
OH-58D	44	112
OH-58D (Armed)	110	112
UH-60 A/L/HH	178	362
CH-47	514	1030

E-72. Aircraft may be capable of carrying more than is indicated on the lists in table E-10 and table E-11, page E-16. However safety, loading procedures, and space limitations play a large part in determining authorized loads for each helicopter. Environmental conditions (high altitude/high temperature decrease max gross weight) and configuration (internal load size/dimensions) constraints affect the ACL for each aircraft.

Table E-10. Typical helicopter load capacities

Type	*Empty Wt Plus Crew & Fuel*	*Max Gross Wt*	*Max Sling Load*
UH-60A	15,000	22,000	8,000
UH-60L	15,000	23,500*	9,000
CH-47D	30,000	50,000*	26,000
*Max gross weight from 8,000 to 9,000 lbs			

Table E-11. Typical planning weights for combat equipment and vehicles

Vehicle/Equipment	*Weight in Pounds*
M998 HMMWV	7,535
M996 TOW HMMWV	8,095
M149 Water Buffalo (Empty) (Loaded)	2,540 6,060
M101A1 ¾ Ton Trailer (Empty) (Loaded)	1,350 2,850
500 Gallon Fuel Drum (Empty) (Full) JP	275 3,625
M102 105mm Howitzer	3,360
M119 105mm Howitzer	4,000
M114A1 155mm Howitzer	15,200
M198 155mm Howitzer	15,740
M167 Vulcan (Towed)	3,260
A22 Bag (Loaded)	2,200
Conex, Steel, Empty	2,140
Conex, Aluminum, Empty	14,600
Conex (Either) Max Load	6,500
Scamp Crane	1,560
One Mil-Van	4,710
Electronic Shop with Wheels	3,940
Tool Set, Shop with Wheels	3,030
Shop, Portable, Aircraft Maintenance (Empty) (Loaded)	4,220 5,425
M1008 Pick-Up (Empty) (Loaded)	5,900 8,800
JD-550 Dozer	16,800

SECTION II – FIXED-WING AIRCRAFT

CARGO-12 HURON (C, D, T1, AND T2 MODELS)

E-73. The C-12 provides high speed air movement capability.

DESCRIPTION

E-74. The C-12 is a twin-engine, turboprop, FW aircraft. Many different C-12 models are fielded. The C-12C and D1 have PT6A-41 engines; the C-12D2, T1, and T2 have PT6A-42 engines. Aircraft can normally carry eight passengers and a crew of two. All models have an aft passenger door, and all (except the C-12C) have a separate cargo door. Table E-12, page E-17, outlines C-12 specifications.

Table E-12. C-12 specifications	
Length	43 ft 10 in
Height	15 ft 5 in
Wingspan	C-12C: 54 ft 6 in, C-12D: 55 ft 6.5 in, C-12T1/T2: 55 ft 6.5 in
Max gross weight at takeoff	13,500 lbs. (C and D1 models), 14,000 lbs.(D2, T1, and T2 models)
Cargo door dimensions	C-12C not installed C-12D and C-12T1/T2 52 in x 52 in
Cruise airspeed	Max 260 kts indicated airspeed, varies with conditions
Ceiling	Max 31,000 MSL, varies with conditions
Range	Varies; for example, 386-gallon main fuel yields 960 NM & 4.5 hours endurance (standard day, zero wind, cruise pressure altitude of 26,000 ft, 1,700 RPM): same conditions w/544-gallon full main & auxiliary fuel yield more than 1,600 NM & 7 hrs endurance
Crew	2 pilots

CAPABILITIES

E-75. The C-12 provides the following:

- Transport of up to eight personnel.
- Communication equipment capable of supporting key passengers.
- Light cargo transport capability.

ARMAMENT SYSTEMS

E-76. The C-12 is unarmed.

COMMUNICATIONS

E-77. The C-12 has the following communication systems, depending on the model:

- The AN/ARC-164 (C, D1, and D2) provides two-way voice communications in the 225 to 399.975 MHz range for a normal range of 50 miles.
- The UHF-20B (C and D1) provides VHF-AM communications in the 116- to 151.975-MHz frequency range for a normal range of 50 miles.
- The 718U HF command set (C and D1) provides high frequency communications in the frequency range of 2 to 29.999 MHz.
- The AN/ARC-186 (C and D1) provides VHF-AM/FM communications.
- The AN/ARC-210 (V) (T1 and T2) provides multifrequency communications in the 30 to 88 FM band, 108 to 136 AM band, 136 to 156 FM band, 156 to 174 FM maritime band, and 225 to 400 AM/FM Have Quick and SATCOM bands.
- The VHF-22C (D2, T1, and T2) provides VHF communications.
- The KHF 950 (D2, T1, and T2) provides high frequency, long-range communications.

NAVIGATION SYSTEMS

E-78. The C-12 has the following navigation systems:

- The KLN-90B GPS provides GPS navigation.
- Two very high frequency omnidirectional range instrument landing system receiver (VIR)-30s are installed, one without marker beacon capability for instrument navigation.
- The KR 87 ADF provides automatic direction finding capability with AM transmitters.

- The DME-40 provides distance-measuring equipment capability.
- The AP-106 autopilot system works with other navigation equipment to fly the aircraft en route.

LIMITATIONS

E-79. The C-12 has no self-defense protection system and is not normally flown at terrain-flight altitudes.

CARGO-23 SUPER SHERPA (B AND B+)

E-80. The C-23B or B+ supports theater aviation air movement needs.

DESCRIPTION

E-81. The C-23B Super Sherpa is a twin turboprop, FW aircraft. Its rectangular-shaped cabin readily accommodates palletized cargo; up to 500 pounds of additional baggage can be stored in a nose compartment. The C-23B has a crew of three. Table E-13 outlines C-23B specifications.

Table E-13. C-23B/B+ specifications

Length	58 ft ½ in
Wingspan	74 ft 8 in
Height	16 ft 3 in
Cabin dimensions	29 ft long x 5 ft 6 in wide x 6 ft 6 in high
Maximum payload	7,100 lbs
Maximum gross wt	25,500 lbs
Maximum airspeed	190 kts
Range	More than 1,000 miles (varies w/environmental/mission conditions)
Typical mission range	770 miles w/5,000-lb payload

CAPABILITIES

E-82. The C-23 provides transport to the following:

- Up to 30 seated passengers.
- 27 paratroopers.
- 18 litters and 2 medical attendants.
- Palletized cargo.

ARMAMENT SYSTEMS

E-83. The C-23B is an unarmed aircraft.

COMMUNICATIONS

E-84. The C-23B and B+ lack SINCGARS, Have Quick, and HF capability. The two organic AN/ARC-182(V) radios operate in the 30 to 399.975 MHz frequency ranges.

NAVIGATION SYSTEMS

E-85. The C-23B has the following navigation systems:

- Two VIR-32A VHF navigation receivers with DME 42 and instrument landing system/glide slope.
- Two radio magnetic indicator-36s.
- One ADF-60A ADF.
- Two electronic horizontal situation indicator-74s.

- One transmittal data relay-90 transponder.
- One AN/APX-100(V) transponder.

C-23B Limitations

E-86. The C-23B is not pressurized; therefore, aircrew members and passengers require oxygen for sustained flights above 10,000 feet.

E-87. The narrow cabin will not permit internal loading of Army vehicles such as the HMMWV.

Appendix F
Rules of Engagement

ROE are directives issued by competent military authority delineating circumstances and limitations under which U.S. forces initiate/continue combat engagement with other forces.

SECTION I – GENERAL

F-1. In a general war between two uniformed, similarly-equipped opponents, complexity of ROE is normally low. The Soldier in contact with the enemy is usually instructed on the priority of target classes to engage, with restricted targets delineated by the Laws of War. In the same conflict, however, Soldiers performing support missions may find their ROE are more complex, reflecting ROE more often associated with stability operations, rather than those associated with direct combat.

F-2. ROE must be clear. Soldiers operating with confusing or uncertain guidance can compromise the mission.

F-3. Despite similarities existing between operations, each has its own ROE. These rules are generally delineated in the OPLAN ROE annex (figure F-1, page F-2); however, based on changing circumstances, they may be further refined in the OPORD. For continuing operations, any further changes are specified in follow-on FRAGOs. The overall commander approves these rules with advice from the Staff Judge Advocate, S-9, political advisor, and others as required.

LAW OF WAR

F-4. International treaties signed by the U.S., such as the Geneva Convention, and customary laws found in federal documents or judicial decisions form the basis of the Law of War to which Soldiers are expected to adhere. The purpose of establishing and abiding by Laws of War is to—

- Protect both combatants and noncombatants from unnecessary suffering.
- Safeguard certain fundamental human rights of persons who fall into the hands of the enemy, particularly prisoners of war, the wounded and sick, and civilians.
- Facilitate the restoration of peace.

F-5. ROE, as with any military order, cannot violate the Laws of War. Those that do are illegal and are not to be followed. Refer to FM 27-10 for additional information.

Copy ____ of _____ copies
HQs, TF 1-19
Camp Deployed, Any Country
Date/Time Group

ANNEX E (ROE) to 1-19 TF OPORD 01-01

References: No change.

1. ROE.
ROE will be briefed in detail to all Soldiers upon issuance of each 1-19 TF OPLAN/OPORD/FRAGO. The commander will resolve conflicts between ROE and the 1-19 TF OPLAN/OPORD/FRAGO.
Nothing in these rules limits the rights of individual Soldiers to defend themselves or the rights and responsibilities of leaders to defend their units.
ROE follow:
SOLDIERS CARD: You will carry this card at all times.
MISSION: Your mission is to assist in the implementation of and to help ensure compliance with this peacekeeping operation.
SELF DEFENSE:
You have the right to use necessary and proportional force in self-defense.
You will use only the minimum force necessary to defend yourself.
GENERAL RULES:
You will use only the minimum force necessary to accomplish your mission.
You will not harm hostile force/belligerents who want to surrender. Disarm them and turn them over to your superiors.
You will treat everyone, including civilians and detained hostile forces/belligerents, humanely.
You will collect and care for the wounded, whether friend or foe.
You will respect private property. Do not steal. Do not take war trophies.
You will prevent or report to your superiors all suspected violations of the Law of Armed Conflict.
CHALLENGING AND WARNING SHOTS:
If the situation permits, issue a challenge:
English: U.S. Forces! STOP or I WILL FIRE
Local Language #1: U.S. Forces! STOP or I WILL FIRE!
Local Language #2: U.S. Forces! STOP or I WILL FIRE!
If the person fails to halt, you may be authorized by the OSC or by standing orders to fire a warning shot.
OPENING FIRE: You may open fire only if you, friendly forces, or persons or properties under your protection are threatened with deadly force. This means that:
You may open fire against an individual who fires or aims a weapon at, or otherwise demonstrates intent to imminently attack you, friendly forces, or persons or property designated as under your protection.
You may open fire against an individual who plants, throws, or prepares to throw an explosive or incendiary device at, or otherwise demonstrates intent to imminently attack you, friendly forces, or persons or property designated as under your protection.
You may open fire against an individual deliberately driving a vehicle at you, friendly forces, or persons or property designated as under your protection.
You may fire against an individual who attempts to take possession of friendly force weapons, ammunition, or property designated as under your protection if there is no other way to prevent this act.
You may use minimum force, including opening fire, against an individual who unlawfully commits, or is about to commit, an act which endangers life, in circumstances if there is no other way to prevent the act.
MINIMUM FORCE: If you have to open fire, you must:
Fire only aimed shots.
Fire no more rounds than necessary.
Take all reasonable efforts to avoid unnecessary destruction of property.
Stop firing as soon as the situation is resolved.
Refrain from intentional attack on civilians, or property that is exclusively civilian or religious in character unless the property is being used for military purposes or engagement is authorized by your commander.

Figure F-1. Example rules of engagement operation plan/operation order/fragmentary order annex

SECTION II – FORCE-PROTECTION LEVELS

F-6. Department of Defense Directive (DODD) 2000.12 sets out DOD Antiterrorism/Force Protection Program responsibilities. DODD 0-2000.12H establishes guidance for force-protection levels. Department of Defense Instruction (DODI) 2000.16 sets out responsibilities for establishing force-protection levels. Full references can be downloaded from http://www.dtic.mil/whs/directives.

F-7. The graduated series of force-protection conditions range from force-protection conditions normal to force protection conditions delta. Table F-1 provides force-protection measures.

Table F-1. Force protection measures

Personnel		*Vehicle*		*Weapon*		*Base Camp*	
O	Soft cap	H	2 veh, 2 pax/veh	W	Wpn and ammo in arms room	A	One roving patrol, towers & fighting positions unmanned, QRF on 2-hour recall
I	Kevlar, load-bearing equipment (LBE), weapon	I	3 veh, 2 pax/veh, M16+ per veh, commo check every hour	X	Wpn carried, magazine in pouch	B	2 pax in towers, 1 roving patrol, QRF 1-hour recall, commo check every hour
II	Kevlar, LBE, weapon, body armor	J	4 veh, 2 pax/veh, M16+ per veh, crew-served wpn, commo check every 30 min	Y	Wpn carried, magazine in wpn, no rounds chambered	C	2 pax in towers, 2 roving patrols, QRF 30-min recall, commo check every 30-min
III	Kevlar, LBE, weapon, body armor, mask	K	4 veh, 2 pax/veh, M16+ per veh, crew-served wpn, continuous commo, military police escort, line traffic coordinator permission	Z	Wpn carried, magazine in wpn, rounds chambered, wpn on safe	D	2 pax in towers, 2 roving patrols, all fighting positions manned; pull in OPs, CPs, & remote sites, QRF at REDCON 1, continuous comunication

F-8. The four-force protection conditions above normal are described in the following paragraphs.

ALPHA

F-9. These conditions apply when there is a general threat of possible terrorist activity against personnel and facilities, the nature and extent of which are unpredictable, and circumstances do not justify full implementation of force protection conditions BRAVO measures. The measures in these force protection conditions must be capable of being maintained indefinitely.

BRAVO

F-10. These conditions apply when an increased and more predictable threat of terrorist activity exists. The measures in these force-protection conditions must be capable of being maintained for weeks without causing undue hardship, affecting operational capability, and aggravating relations with local authorities.

CHARLIE

F-11. These conditions apply when an incident occurs or intelligence is received indicating some form of terrorist action against personnel and facilities is imminent. Implementation of measures in these force-protection conditions for more than a short period may create a hardship and affect peacetime activities of the unit and its personnel.

DELTA

F-12. These conditions apply in the immediate area where a terrorist attack has occurred or when intelligence has been received that terrorist action against a specific location or person is likely. Normally, these force-protection conditions are declared as a localized condition.

SECTION III – RULES OF ENGAGEMENT REHEARSALS

F-13. ROE cards are excellent reminders, however to ensure ROE are understood, situational training exercises are essential. For example, an ROE may be "fire only in self-defense." But when does a Soldier know he or she is really being fired upon? Is it possible the shooter is just some inebriated person in the crowd shooting into the air? If an aircrew sees tracers going by, does it mean personnel on the ground saw the aircraft and are firing at it or is it just celebratory fire?

F-14. This manual will not debate the correct answer to the above; it only points out that Soldiers need clear examples of situations to ensure they fully understand the ROE. An exercise to demonstrate the above would require role players, the training unit, observer-controllers, and a range safety plan.

F-15. ROE situations should be rehearsed in detail before deploying or executing a mission. No situation should occur in which personnel are unsure whether they should use force and what types of force, to include deadly force, are warranted.

Appendix G

Brigade Command Post Layout

This appendix provides a suggested layout for CAB command posts and individual cells. These designs are intended to standardize CP layouts as described in FMI 5-0.1. Refer to FMI 5-0.1 for additional information.

SECTION I – OVERVIEW

G-1. Many design considerations affect CP effectiveness. At a minimum, CP cells and staff elements should be positioned to facilitate communication and coordination. Other design considerations include—

- Ease of information flow.
- User interface with communication systems.
- Positioning information displays for ease of use.
- Integrating complementary information on maps and displays.
- Adequate workspace for the staff and commander.
- Ease of displacement (set-up, tear-down, and mobility).

G-2. A well-designed CP integrates command and staff efforts. Accomplishing this requires matching CP personnel, equipment, information systems, and procedures with its internal layout and utilities. Organizing the CP into functional and integrating cells promotes efficiency and coordination.

G-3. Standardization increases efficiency and eases CP personnel training. Commanders should develop detailed SOPs for all aspects of CP operations. These SOPs should be revised throughout training activities.

FUNCTIONAL CELLS

G-4. Functional cells are organized by WFF with the addition of the C4OPS cell.

INTELLIGENCE

G-5. The intelligence cell coordinates activities and systems that facilitate understanding the enemy, terrain, weather, and civil considerations. This includes tasks associated with IPB and ISR. The unit's S-2 leads this cell.

MOVEMENT AND MANEUVER

G-6. The movement and maneuver cell coordinates activities and systems that move forces to achieve a position of advantage in relation to the enemy. This includes tasks associated with employing forces in combination with direct fire or fire potential (maneuver), force projection (movement), mobility, and countermobility. The movement and maneuver cell may also form the base of the current operations cell. The unit's S-3 or assistant S-3 leads this cell.

FIRE SUPPORT

G-7. The fire support cell coordinates activities and systems that provide collective and coordinated use of Army indirect fires and joint fires. This includes tasks associated with targeting and the targeting process.

The fire support cell integrates lethal and nonlethal fires, including IO, through the targeting process. The unit's fire support coordinator leads this cell.

Protection

G-8. The protection cell coordinates activities and systems that preserve the force. This includes protecting personnel, physical assets, and information of the U.S. and multinational partners. Other tasks are found in FMI 5-0.1. Commanders normally select this cell's leader from among the air and missile defense coordinator, chemical officer, engineer coordinator, and provost marshal.

Sustainment

G-9. The sustainment cell coordinates activities and systems that provide support and services to ensure freedom of action, extend operational reach, and prolong endurance. Other tasks are found in FMI 5-0.1. The commander normally selects either the S-1 or S-4 as the cell leader.

Command, Control, Communications, and Computer Operations

G-10. The C4OPS cell coordinates activities and systems that provide support to continuous and assured communications. This includes tasks associated with C4OPS, network operations, and information systems support to information management. The unit's S-6 leads this cell.

INTEGRATING CELLS

G-11. Integrating cells group personnel and equipment to integrate functional cell activities. CPs normally include current operations, future operations, and plans cell. The plans cell is normally located in the main CP. The current operations and future operations cells are normally located at the TAC CP.

Current Operations

G-12. The current operations cell is responsible for assessing the current situation while regulating forces and WFFs in accordance with the commander's intent. Normally, all staff sections are represented in the current operations cell. The unit's S-3, or assistant S-3, leads this cell. Personnel in the movement and maneuver cell are also normally members of the current operations cell.

G-13. Staff representatives in the current operations cell actively assist subordinate units. They provide them information, synchronize their activities, and coordinate their support requests. The current operations cell solves problems and acts within the authority delegated by the commander. It also performs short-range planning using MDMP in a time-constrained environment or makes decision and resynchronizes operations as described in FMI 5-0.1.

Future Operations

G-14. The future operations cell is responsible for planning and assessing operations for the mid-range time horizon. This includes preparing branches. Corp and Army service component commands have a FUOPS cell. Battalion through division headquarters are not resources for one; the plans and current operations cells share its responsibility. The future operations use MDMP or MDMP in a time-constrained environment to develop plans and orders. The cell consists of a core group of planners led by the assistant S-3. All staff sections assist as required.

Plans

G-15. The plans cell is responsible for planning operations for the mid- to long-range time horizons. It develops plans, orders, branches, and sequels. This cell is also responsible for long-range assessment of an operation's progress. It consists of a core group of planners and analysts led by the S-5. All staff sections assist as required.

SECTION II – COMBAT AVIATION BRIGADE MAIN COMMAND POST

G-16. The main command post is a command and control facility that contains the portion of the unit headquarters in which the majority of planning, analysis, and coordination occurs (FMI 5-0.1). The main CP includes representatives of all staff sections. It is larger in size and personnel and less mobile than the TAC CP. The main CP controls current operations when the TAC CP cannot or is not employed. Figure G-1, figure G-2 (page G-4), figures G-3 and G-4 (page G-5), figure G-5 and G-6 (page G-6), and figures G-7 and G-8 (page G-7) illustrate a typical layout for a CAB main CP.

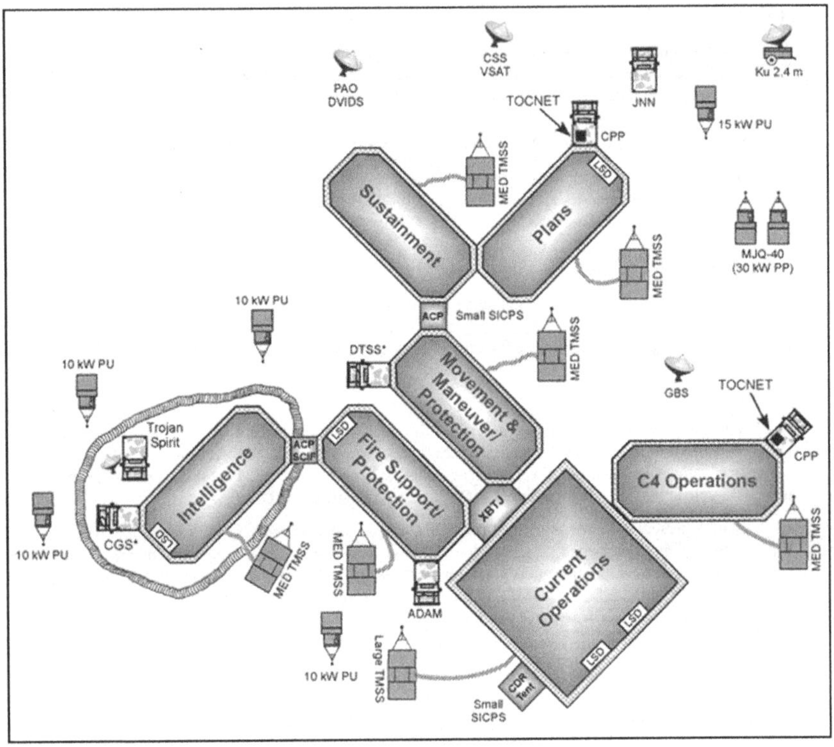

Figure G-1. Main command post

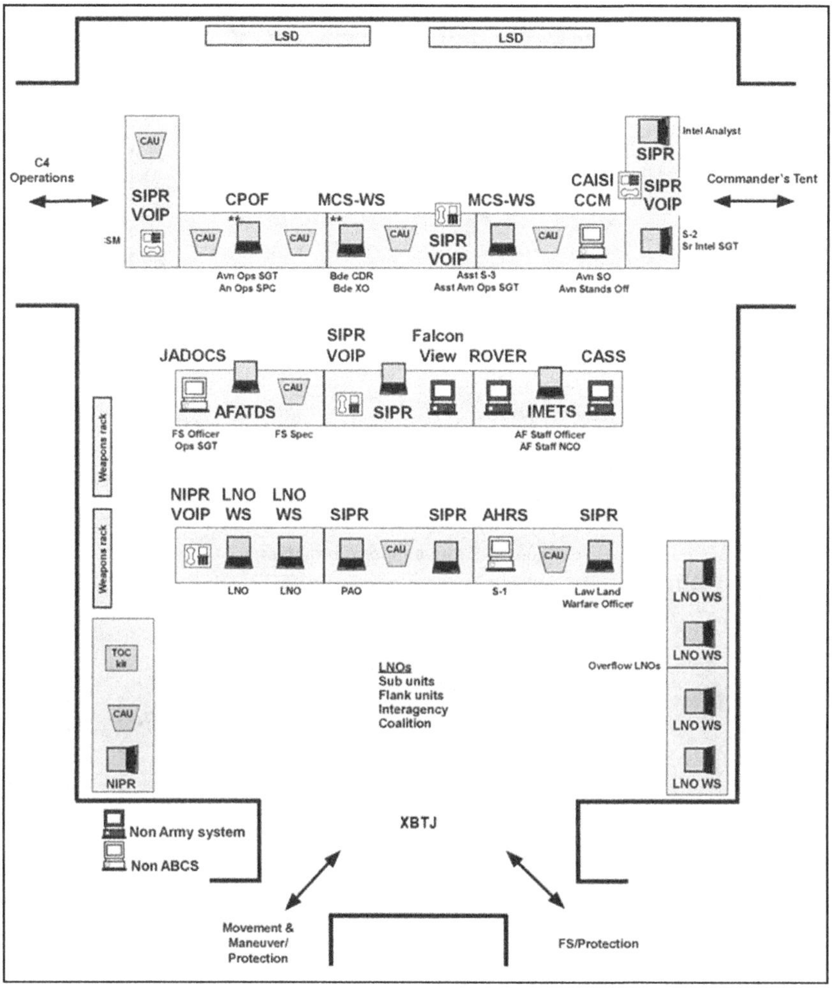

Figure G-2. Current operations

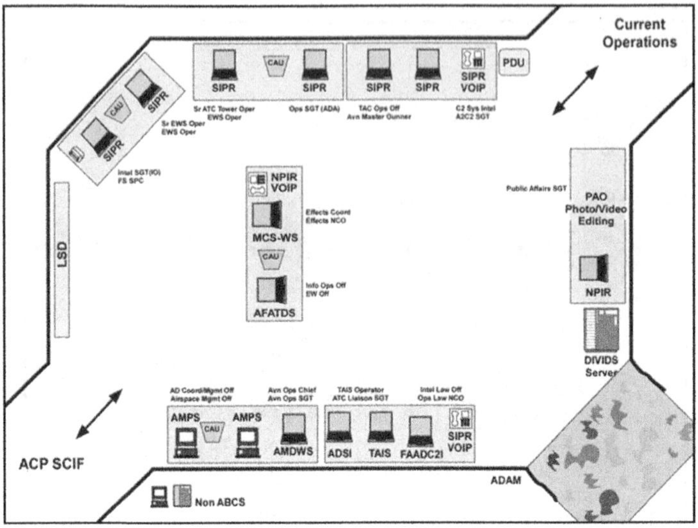

Figure G-3. Fire support/protection

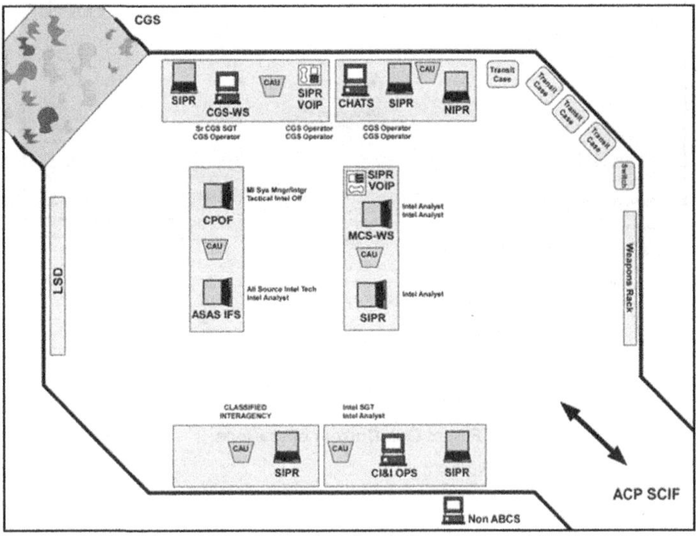

Figure G-4. Intelligence

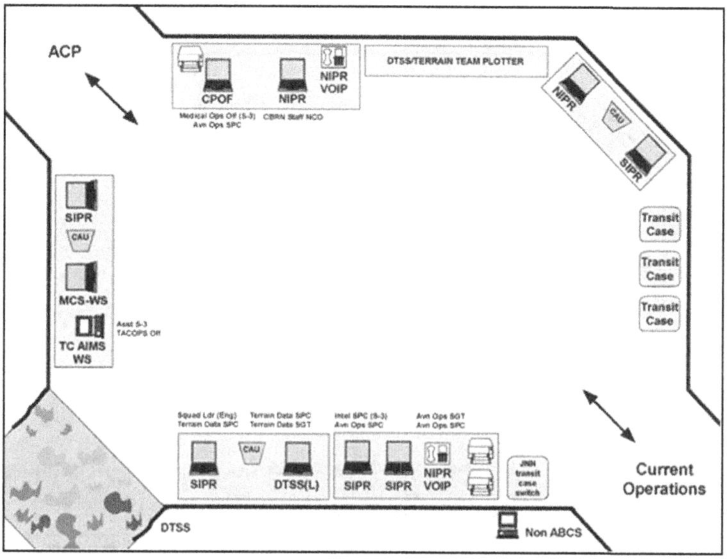

Figure G-5. Movement and maneuver/protection

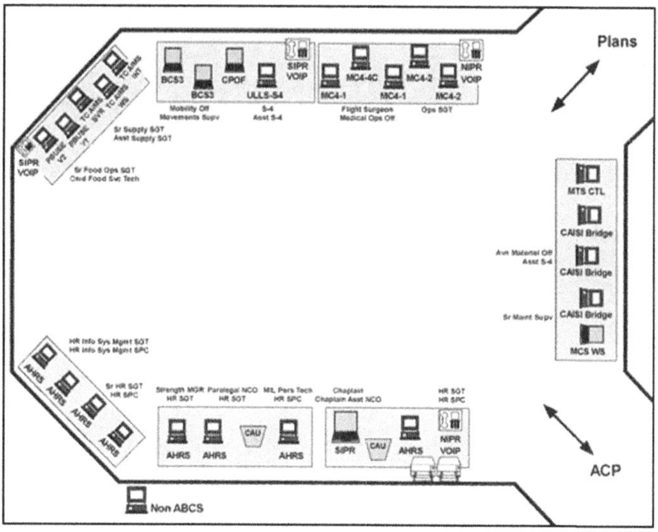

Figure G-6. Sustainment

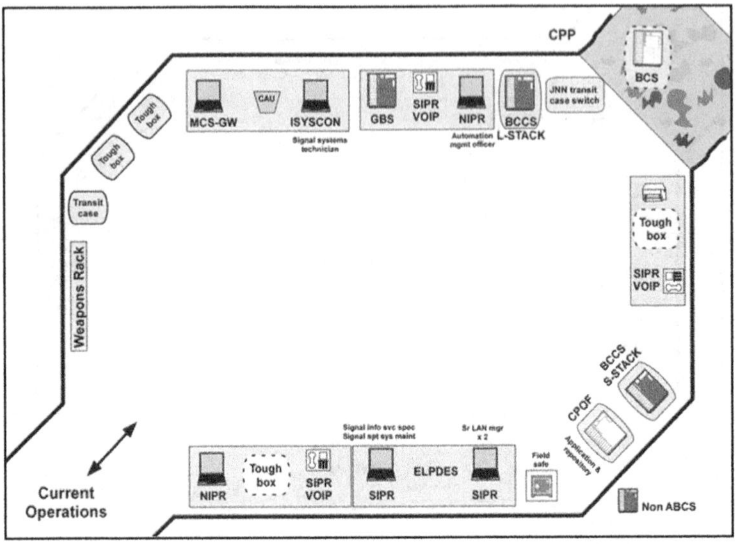

Figure G-7. Command, control, communications, and computers operations

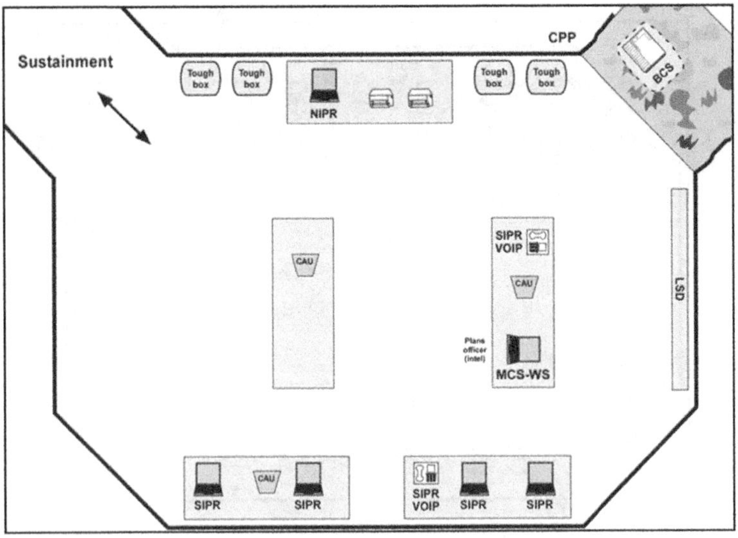

Figure G-8. Plans

SECTION III – COMBAT AVIATION BRIGADE TACTICAL COMMAND POST

G-17. The tactical command post is a command and control facility containing a tailored portion of a unit headquarters designed to control current operations (FMI 5-0.1). The TAC CP is fully mobile. As a rule, it includes only Soldiers and equipment essential to control current operations. The TAC CP relies on the main CP for planning, detailed analysis, and coordination. The S-3 usually leads the TAC CP. Figures G-9 and G-10 and figure G-11 (page G-9) illustrate a typical TAC CP layout.

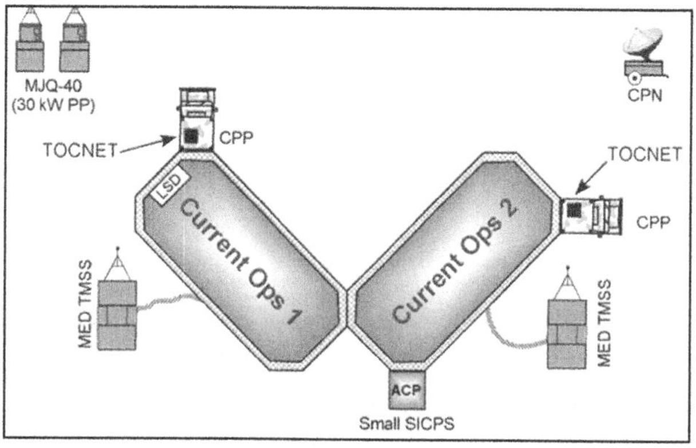

Figure G-9. Tactical command post

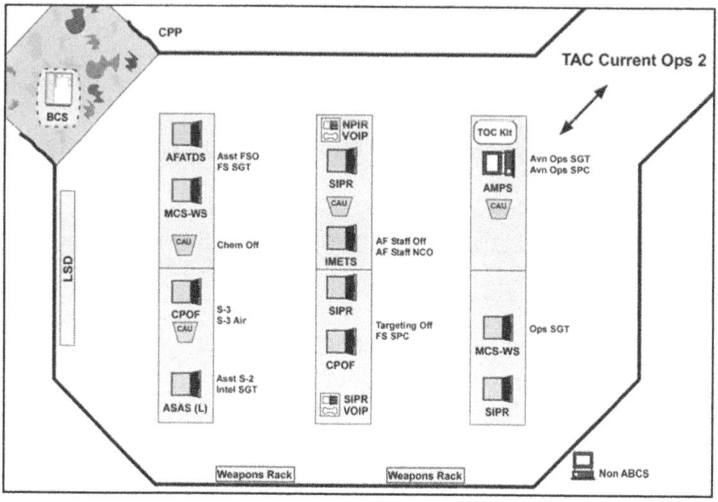

Figure G-10. Current operations 1

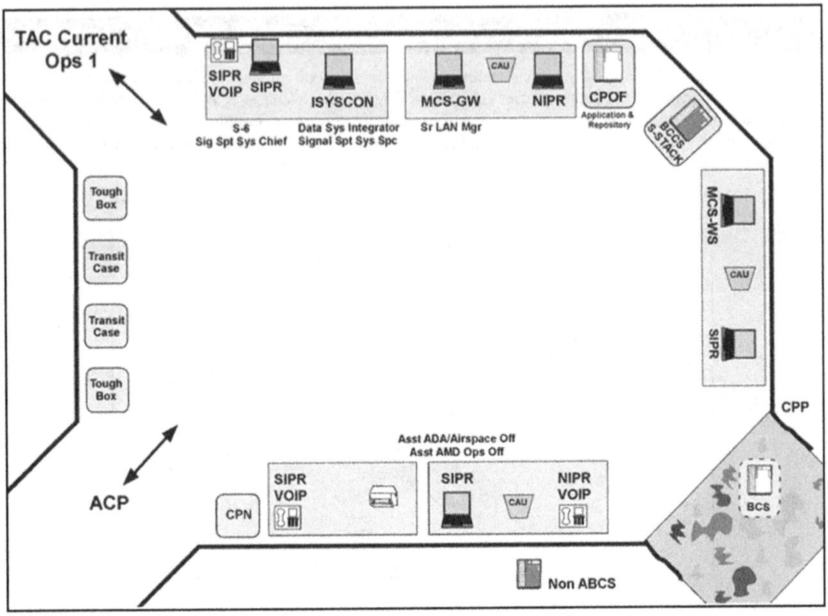

Figure G-11. Current operations 2

Glossary

1SG	first sergeant
A&L	administrative and logistics
AC2	airspace command and control
A2C2S	Army airborne command and control system
AA	assembly area
AAMC	air ambulance medical company
AATF	air assault task force
ABCS	Army battle command system
ABTF	aviation battle task force
ACE	analysis and control element
ACL	allowable cabin load
ACM	airspace coordinating measure
ACO	air control order
ACS	air cavalry squadron
AD	air defense
ADAM	air defense and airspace managment
AEB	aviation expeditionary brigade
AFATDS	advanced field artillery tactical data system
AH	attack helicopter
AHB	attack helicopter battalion
AHC	assault helicopter company
AI	air interdiction
AKO	Army Knowledge Online
ALE	automatic link establishment
ALO	air liaison officer
ALSE	aviation life support equipment
AM	amplitude modulated
AMC	air mission commander
AMDWS	air and missile defense work station
AMO	aviation materiel officer
AMPS	aviation mission planning system
AMSS	Army materiel status system
ANCD	automated network control device
AO	area of operations
APOD	aerial port of debarkation
AR	Army regulation
ARB	attack reconnaissance battalion
ARC	attack reconnaissance company

ARFORGEN	Army force generation
ARNG	Army National Guard
ARP	airframe repair platoon
ARS	attack reconnaissance squadron
ART	attack reconnaissance troop
ASAS	all source analysis system
ASAS-L	all source analysis system-light
ASB	aviation support battalion
ASE	aircraft survivability equipment
ASL	authorized stockage list
ATCCS	Army Tactical Command and Control System
ATHP	ammunition transfer holding point
ATO	air tasking order
ATP	ammunition transfer point
ATS	air traffic services`
ATX	aviation training exercise
AVCATT	aviation combined arms tactical trainer
AWACS	airborne warning and control system
BAE	brigade aviation element
BAS	battlefield automated system
BCOTM	battle command on the move
BCS-3	battle command sustainment support system
BCT	brigade combat team
BDA	battle damage assessment
BDAR	battle damage assessment and repair
BFT	Blue Force Tracker
BOLT	brigade operational law team
BP	battle position
BSA	brigade support area
BSS	brigade surgeon section
C2	command and control
C4OPS	command, control, communications, and computer operations
CAB	combat aviation brigade
CAC	command aviation company
CAS	close air support
CASEVAC	casualty evacuation
CBRN	chemical, biological, radiological, and nuclear
CBRNE	chemical, biological, radiological, nuclear, and high yeild explosives
CCA	close combat attack
CCIR	commander's critical information requirement

CEF	contingency expeditionary force
CGS	common ground station
CH	cargo helicopter
CIC	combat information center
CMO	civil-military operations
COA	course of action
COMSEC	communications security
CONUS	continental United States
COP	common operational picture
CP	command post
CRM	composite risk management
CRP	component repair platoon
CSAR	combat search and rescue
CSM	command sergeant major
CSR	controlled supply rate
CSSAMO	Combat Service Support Automation Management Office
CSSCS	combat service support control system
CTA	common table of allowances
CTC	combat training center
CTP	common tactical picture
DA	Department of the Army
DA Pam	Department of the Army pamphlet
DART	downed aircraft recovery team
DEF	deployment expeditionary force
DOD	Department of Defense
DOD FLIP	Department of Defense flight information publication
DODD	Department of Defense directive
DODI	Department of Defense instruction
DOS	day of supply
DP	decisive point
DS	direct support
DSN	defense switch network
DTSS	digital topographical support system
EA	engagement area
EAB	echelons above brigade
EAC	echelons above corps
EEFI	essential elements of friendly information
EGI	embedded global positioning system/inertial navigation system
EPLRS	enhanced position location reporting system
ERFS	extended range fuel system

ESSS	external stores support system
EUH	electronic utility helicopter
EW	electronic warfare
FA	field artillery
FAA	forward assembly area
FAC	forward air controller
FARE	forward area refueling equipment
FARP	forward arming and refueling point
FBCB2	Force XXI Battle Command Brigade and Below
FCR	fire control radar
FLOT	forward line of own troops
FM	field manual, frequency modulated
FMI	field manual interim
FP	firing position
FRAGO	fragmentary order
FRG	family readiness group
FRIES	fast-rope insertion/extraction system
FS	fire support
FSC	forward support company
FSCM	fire support coordinating measure
FSCOORD	fire support coordinator
FSE	fire support element
FSMT	forward support medical evacuation team
FW	fixed-wing
G-3	Assistant Chief of Staff-Operations
G-4	Assistant Chief of Staff-Logistics
G-6	Assistant Chief of Staff-Command, Control, Communications and Computer Operations (C4 Ops)
GCCS	global command and control system
GCCS-A	global command and control system-Army
GIG	global information grid
GPS	global positioning system
GSAB	general support aviation battalion
GS	general support
GSE	ground support equipment
HA	holding area
HF	high frequency
HHC	headquarters and headquarters company
HMMWV	high mobility multi-purpose wheeled vehicle
HPT	high-payoff target
HSC	headquarters and support company

HSS	health support service
HUD	heads-up display
HVT	high-value target
HvyHC	heavy helicopter company
IDM	improved data modem
IFF	Identification, friend or foe
IFR	instrument flight rules
IM	information management
IMC	instrument meteorological condition
IMDC	isolated, missing, detained, or captured
IMETS	integrated meteorological system
INC	internet controller
INS	inertial navigation system
INTREP	intelligence report
IO	information operations
IP	instructor pilot
IPB	intelligence preparation of the battlefield
IR	infrared
ISR	intelligence, surveillance, and reconnaissance
ITV	in-transit visibility
JAAT	joint air attack team
JCDB	joint common database
JFC	joint force commander
JIM	joint, interagency, and multinational
JNN	joint network node
JNTC	joint network tactical node
JP	joint publication
JPRC	joint personnel recovery center
J-SEAD	joint suppression of enemy air defense
JSTARS	joint surveillance target attack radar system
JTF	joint task force
JVMF	joint variable message format
LAN	local area network
LBE	load-bearing equipment
LCOP	logistics common operational picture
LD	line of departure
LNO	liaison officer
LOC	line of communications
LOGPAC	logistics package
LOS	line of sight

LPB	logistics preparation of the battlefield
LRU	line replaceable unit
LSD	large screen display
LUH	light utility helicopter
LZ	landing zone
MAC	maintenance allocation chart
MCO	major combat operations
MCS	maneuver control system
MCS-L	maneuver control system-light
MDMP	military decisionmaking process
MEDEVAC	medical evacuation
METL	mission essential task list
METT-TC	mission, enemy, terrain and weather, troops and support available, time available, and civil considerations
MMC	materiel management center
MMS	mast-mounted sight
MO	maintenance officer
MOC	medical operations cell
MOPP	mission-oriented protective posture
MOS	military occupational specialty
MP	maintenance test pilot
MSE	mobile subscriber equipment
MSO	mission staging operation
MSR	main supply route
MST	maintenance support team
MTF	medical treatment facility
MTOE	modified table of organization and equipment
MTS	movement tracking system
NAI	named area of interest
NATO	North Atlantic Treaty Organization
NAVAID	navigational aid
NCO	noncommissioned officer
NCS	net control station
NGA	National Geospatial-Intelligence Agency
NIPR	nonsecure internet protocol router
NIPRNET	nonsecure internet protocol router network
NLOS	non-line of sight
NOE	nap-of-the-earth
NSC	network support company
NSFS	naval surface fire support
NTDR	near term digital radio

NVD	night vision device
NVG	night vision goggle
O&I	operations and intelligence
OCONUS	outside the continental United States
OE	operational environment
OH	observation helicopter
OP	observation post
OPCON	operational control
OPLAN	operation plan
OPORD	operation order
OPSEC	operations security
OPTEMPO	operating tempo
OSC	on-scene commander
PAO	public affairs officer
PIR	priority intelligence requirement
PLL	prescribed load list
PMCS	preventive maintenance checks and services
PNVS	pilot night vision sensor
POC	point of contact
POD	port of debarkation
POE	port of embarkation
POL	petroleum, oils, and lubricants
PP	passage point
PR	personnel recovery
PRCC	personnel recovery coordination cell
PRO	personnel recovery officer
PZ	pickup zone
PZCO	pickup zone control officer
QA	quality assurance
QRF	quick reaction force
R4	redeployment, reintegration, reconstitution, and retraining
RC	reserve component
REDCON	readiness condition
REF	ready expeditionary force
RESCORT	rescue escort
RMC	rescue mission commander
ROE	rules of engagement
ROI	rule of interaction
RP	release point
RSOI	reception, staging, onward movement, and integration

RSR	required supply rate
RWS	remote workstation
S&S BN	security and support battalion
S-1	personnel staff officer
S-2	intelligence staff officer
S-3	operations staff officer
S-4	logistics staff officer
S-5	plans staff officer
S-6	command, control, communications and computer operations (C4 Ops) staff officer
S-9	civil affairs staff officer
SA	situational awareness
SAL	semi-active laser
SAMS-1	standard Army maintenance system-level 1
SAMS-2	standard Army maintenance system-level 2
SAMS-E	standard Army maintenance system-enhanced
SAR	search and rescue
SARSS-1	standard Army retail supply system-level 1
SARSS-2A/C	standard Army retail supply system-level 2A/C
SARSS-Gateway	standard Army retail supply system-gateway
SARSS-O	standard Army retail supply system-objective
SATCOM	satellite communications
SDDCTEA	Surface Deployment and Distribution Command Transportation Engineering Agency
SEAD	suppression of enemy air defense
SEN	small extension node
SERE	survival, evasion, resistance, and escape
SICPS	standardized integrated command post system
SINCGARS	single-channel ground and airborne radio system
SINCGARS-SIP	single-channel ground and airborne radio system-system improvement program
SIPR	secret internet protocol router
SIPRNET	SECRET Internet Protocol Router Network
SITREP	situation report
SMART-T	Secure Mobile Anti-Jam Reliable Tactical-Terminal
SME	subject matter expert
SO	safety officer
SOI	signal operation instructions
SOP	standing operating procedure
SP	standardization instructor pilot
SPBS-R	standard property book system-redesign

SPIES	special patrol infiltration/exfiltration system
SPINS	special instructions
SPO	support operations officer
SPOD	sea port of debarkation
SPOTREP	spot report
SSA	supply support area
STAMIS	Standard Army Management Information System
STANAG	Standardization Agreement
STATREP	status report
SU	situational understanding
TA	target acquisition
TAC CP	tactical command post
TACLAN	tactical local area network
TACON	tactical control
TACOPS	tactical operations
TACP	tactical air control party
TACSOP	tactical standing operating procedures
TADS	target acquisition designation sight
TAGS	theater air-ground system
TAIS	tactical airspace integration system
TAV	total asset visibility
TC	training circular
TCF	tactical combat force
TDMA	time division multiple access
TF	task force
TI	tactical internet
TIS	thermal imaging system
TM	technical manual
TOA	transfer of authority
TOD	time of day
TOE	table of organization and equipment
TPFDD	time-phased force and deployment data
TTP	tactics, techniques, and procedures
UAS	unmanned aircraft system
UBL	unit basic load
UH	utility helicopter
UHF	ultra high frequency
ULLS	unit level logistics system
ULLS-A	unit level logistics system-aviation
ULLS-G	unit level logistics system-ground

ULLS-S4	unit level logistics system-supply
UMO	unit movement officer
UMT	unit ministry team
USAF	United States Air Force
USACRC	United States Army Combat Readiness Center
UTO	unit task order
VHF	very high frequency
VIR	very high frequency omnidirectional range instrument landing system receiver
VoIP	voice over internet protocol
VTC	video teleconference
WAN	wide area network
WARNO	warning order
WFF	warfighting function
WIN-T	warfighter information network-tactical
WMD	weapons of mass destruction
XO	executive officer

References

These publications are sources for additional information on the topics in this FM. Most JPs are found online at http://www.dtic.mil/doctrine/jpreferencepubs.htm. Most Army doctrinal publications are found online at http://www.army.mil/usapa/doctrine/Active_FM.html. Most FAA publications are found online at http://www.faa.gov/regulations_policies/. Aeronautical information manual can be found at http://www.faa.gov/ATpubs/AIM/

SOURCES USED

These are the sources quoted or paraphrased in this publication.

ARMY PUBLICATION

AR 25-2, *Information Assurance*, 24 October 2007.

AR 95-1, *Flight Regulations*. 3 February 2006.

AR 200-1, *Environmental Protection and Enhancement*, 28 August 2007.

AR 360-1, *The Army Public Affairs Program*, 15 September 2000.

AR 385-10, *The Army Safety Program*, 23 August 2007.

AR 700-138, *Army Logistics Readiness and Sustainability*, 26 February 2004.

AR 750-1, *Army Materiel Maintenance Policy*, 20 September 2007.

AR 750-6, *Army Equipment Safety and Maintenance Notification System*, 13 October 2006.

CTA 8-100, *Army Medical Department Expendable/Durable Items*, 17 December 2004.

CTA 50-900. *Clothing and Individual Equipment*. 31 July 2003.

DA Pam 385-40, *Army Accident Investigation and Reporting*, 1 November 1994.

DA Pam 385-64, *Ammunition and Explosives Safety Standards. 15 December 1999.*

DA Pam 738-751. *Functional Users Manual for the Army Maintenance Management System—Aviation (TAMMS-A),* 15 March 1999.

DA Pam 750-8, *The Army Maintenance Management System (TAMMS) User Manual,* 22 August 2005.

DODD 2000.12, *DOD Antiterrorism (AT) Program*, 18 August 2003.

DODI 2000.16, *DOD Antiterrorism (AT) Program*, 2 October 2006.

EM 0126, *TM 1-1520-Longbow/Apache, Interactive Electronic Technical Manual (ETM) for Longbow/Apache*, 27 December 2005.

EM 0253. *TM 55-1520-241-S, Interactive Electronic Technical Manual (ETM) for Preparation for Shipment of CH-47 Helicopter*, 30 June 2004.

FM 1-02, *Operational Terms and Graphics*, 21 September 2004.

FM 1-113, *Utility and Cargo Helicopter Opertions*, 12 September 1997.

FM 3-0, *Operations*, 14 June 2001.

FM 3-04.15, *Multi-Service Tactics Techniques, and Procedures for the Tactical Employment of Unmanned Aircraft Systems*, 3 August 2006.

FM 3-04.104, *Tactics, Techniques, and Procedures for Forward Arming and Refueling Point*, 3 August 2006.

FM 3-04.120, *Air Traffic Services Operations*. 16 February 2007.

FM 3-04.126, *Attack Reconnaissance Helicopter Operations*, 16 February 2007.

FM 3-04.301, *Aeromedical Training for Flight Personnel*, 29 September 2000.

FM 3-04.500, *Army Aviation Maintenance*, 23 August 2006.

FM 3-04.513, *Battlefield Recovery and Evacuation of Aircraft*, 27 September 2000.

FM 3-06.1, *Aviation Urban Operations for Multiservice Procedures for Aviation Urban Operations*, 9 July 2005.

FM 3-06.11, *Combined Arms Operations in Urban Terrain*, 28 February 2002.

FM 3-11, *Multiservice Tactics, Techniques, and Procedures for Nuclear, Biological, and Chemical Defense Operations*, 10 March 2003.

FM 3-11.4, *Multiservice Tactics, Techniques, and Procedures for Nuclear, Biological, and Chemical (NBC) Protection*, 2 June 2003.

FM 3-19.1, *Military Police Operations*, 22 March 2001.

FM 3-50.1, *Army Personnel Recovery*, 10 August 2005.

FM 3-52, *Army Airspace Command and Control in a Combat Zone*, 1 August 2002.

FM 4-0, *Combat Service Support*, 29 August 2003.

FM 4-01.011, *Unit Movement Operations*, 31 October 2002.

FM 4-01.30, *Movement Control*, 1 September 2003.

FM 4-01.45.,*Multi-Service Tactics, Techniques, and Procedures for Tactical Convoy Operations*, 24 March 2005.

FM 4-02.21, *Division and Brigade Surgeons' Handbook (Digitized) Tactics, Techniques, and Procedures*, 15 November 2000.

FM 4-30.31, *Recovery and Battle Damage Assessment and Repair*, 19 September 2006.

FM 5-0, *Army Planning and Orders Production*, 20 January 2005.

FM 5-19, *Composite Risk Management*, 21 August 2006.

FM 6-0, *Mission Command: Command and Control of Army Forces*, 11 August 2003.

FM 10-1, *Quartermaster Principles*, 11 August 1994.

FM 11-55, *Mobile Subscriber Equipment (MSE) Operations*, 22 June 1999.

FM 27-10, The *Law of Land Warfare*, 18 July 1956.

FM 71-100-2, *Infantry Division Operations, Tactics, Techniques, and Procedures*, 31 August 1993.

FM 71-100-3, *Air Assault Division Operations for Tactics, Techniques, and Procedures*, 29 October 1996.

FM 100-9, *Reconstitution*, 13 January 1992.

FMI 3-04.155, *Army Unmanned Aircraft System Operations*, 4 April 2006.

FMI 5-0.1, *The Operations Process*, 31 March 2006.

JP 4-0, *Doctrine for Logistic Support of Joint Operations*, 6 April 2000.

JP 4-03, *Joint Bulk Petroleum and Water Doctrine*, 23 May 2003.

MIL-STD-129P, *Military Marking for Shipment and Storage*, 15 December 2002.

TC 1-400, *Brigade Aviation Element Handbook*, 27 April 2006.

TC 38-3, *Guide for Basic Military Preservation and Packing*, 1 December 1999.

TM 1-1520-237-S, *Preparation for Shipment of Army Models*, UH-60A (*NSN 1520-01-035-0266) (EIC:RSA) UH-60L (NSN 1520-01-298-4532) (EIC:RSM) EH-60A (NSN 1520-01-082-0686) (EIC:RSB) HH-60A (NSN 1520-01-459-9468) (EIC:RSN) HH-60L(NSN 1520-01-471-6743) (EIC:RSI) UH-60M (NSN 1520-01-492-6324) (EIC:RSP)HH-60M (NSN 1520-01-515-4615) (EIC:N/A)*, 1 November 2007.

TM 1-1520-248-10, *Operator's Manual for Army OH-58D Helicopter*, 7 February 2007.

TM 1-1520-252-S, *Preparation for Shipment of MH-47E Helicopter* (EIC: RCE), 28 June 1995.

TM 38-250, *Preparing Hazardous Materials for Military Air Shipments,* 12 October 2004.

TM 55-1520-238-S, *Preparation for Shipment Army AH-64A Helicopter (NSN 1520-01-106-9519)*, 30 September 1990.

DOCUMENTS NEEDED

These documents must be available to the intended users of this publication.

DEPARTMENT OF THE ARMY FORMS

DA Form 2028, *Recommended Changes to Publications and Blank Forms.*

DA Form 7566, *Composite Risk Management Worksheet.*

READINGS RECOMMENDED

These sources contain relevant supplemental information.

Aviation Liaison Officer Handbook, USAAWC, July 2001.

AR 385-10, *U.S. Army Explosives Safety Program*, 23 August 2007.

AR 600-55, *The Army Driver and Operator Standardization Program (Selection, Training, Testing, and Licensing)*, 18 June 2007.

AR 710-2, *Supply Policy Below the National Level,* 8 July 2005.

ARTEP 1-111-MTP, *Mission Training Plan for the Aviation Brigades,* 27 October 2005.

ARTEP 1-113-MTP, *Mission Training Plan for the Assault Helicopter Battalion,* 29 December 2005.

ARTEP 1-118-MTP, *General Support Aviation Battalion,* 17 January 2006.

ARTEP 1-126-MTP, *Mission Training Plan for the Attack Reconnaissance Helicopter Battalion/Squadron,* 8 March 2006.

ARTEP 1-500-MTP, *Mission Training Plan for the Aviation Intermediate Maintenance (AVIM) Battalion and Company,* 1 April 2002.

DA Pam 25-30, *Consolidated Index of Army Publications and Blank Forms,* 1 January 2007.

DA Pam 25-33, *User's Guide for Army Publications and Forms,* 15 September 1996.

DA Pam 385-40, *Army Accident Investigation and Reporting,* 1 November 1994.

DA Pam 710-2-1, *Using Unit Supply System (Manual Procedures),* 31 December 1997.

DA Pam 710-2-2, *Supply Support Activity Supply System: Manual Procedures,* 30 September 1998.

EM 0007, *FEDLOG,* 1 December 2007. (S&I, Commander, USAMC Logistics Support Activity, ATTN: AMXLS-MLA, [A. LEWIS] Bldg 5307, Redstone Arsenal, Al 35898-7466.)

FAAO 3120.4L, *Air Traffic Technical Training* , 22 June 2005.

FAAO 7000.5C, *Submissions for Air Traffic Publications,* 18 October 2001.

FAAO 7010.1S, *Air Traffic Control Safety Evaluations and Audits,* 1 October 2005.

FAAO 7110.10S, *Flight Services,* 16 February 2006.

FAAO 7110.65R, *Air Traffic Control,* 16 February 2006.

FAAO 7210.3U, *Facility Operation and Administration,* 16 February 2006.

FAAO 7210.56C, *Air Traffic Quality Assurance,* 15 August 2002.

FAAO 7340.1Z, *Contractions,* 25 October 2007.

FAAO 7350.7B, *Location Identifiers,* 25 October 2007.

FAAO 7400.2F, *Procedures for Handling Airspace Matters,* 16 February 2006.

FAAO 7450.1, *Special Use Airspace Management System,* 21 June 1999.

FAAO 7610.4M, *Special Operations,* 18 January 2007.

FAAO 7900.5, *Surface Weather Observing,* 11 May 2001.

FAAO 7930.2K, *Notices to Airmen (NOTAMS),* 16 February 2006.

FM 1-100, *Army Aviation Operations*, 21 February 1997.

FM 2-0, *Intelligence*, 17 May 2004.

FM 3-04.140, *Helicopter Gunnery*, 14 July 2003.

FM 3-04.300, *Flight Operations Procedures*, 26 April 2004.

FM 3-04.303, *Air Traffic Services Facility Operations, Training, Maintenance, and Standardization,* 3 December 2003.

FM 3-04.508, *Aviation Life Support System Maintenance Management and Training Programs*, 23 April 2004.

FM 3-05.30, *Psychological Operations,* 15 April 2005.

FM 3-05.60, *Army Special Operations Forces Aviation Operations,* 30 October 2007.

FM 3-07, *Stability Operations and Support Operations,* 20 February 2003.

FM 3-07.31, *Peace Operations Multi-Service Tactics, Techniques, and Procedures for Conducting Peace Operations,* 26 October 2003.

FM 3-09.32, *(J-FIRE) Multiservice Procedures for the Joint Application of Firepower*, 29 October 2004.

FM 3-13, *Information Operations: Doctrine, Tactics, Techniques, and Procedures,* 28 November 2003.

FM 3-20-96, *Cavalry Squadron (RSTA),* 20 September 2006.

FM 3-21.20, *The Infantry Battalion,* 13 December 2006.

FM 3-21.38, *Pathfinder Operations,* 25 April 2006.

FM 3-50.3, *Survival, Evasion, and Recovery, Multi-Service Tactics, Techniques, and Procedures for Survival, Evasion, and Recovery,* 20 March 2007.

FM 3-52, *Army Airspace Command and Control in a Combat Zone,* 1 August 2002.

FM 3-52.2, *TAGS Multi-Service Tactics, Techniques, and Procedures for the Theater Air Ground System,* 10 April 2007.

FM 3-90.1, *Tank and Mechanized Infantry Company Team,* 9 December 2002.

FM 3-90.6, *The Brigade Combat Team,* 4 August 2006.

FM 3-100.12, *Risk Management for Multiservices Tactics, Techniques, and Procedures,* 15 February 2001.

FM 3-100.4, *Environmental Considerations in Military Operations,* 15 June 2000.

FM 4-02.2, *Medical Evacuation,* 8 May 2007.

FM 4-20.197, *Multiservice Helicopter Sling Load: Basic Operations and Equipment*, 20 July 2006.

FM 5-34, *Engineer Field Data,* 19 July 2005.

FM 5-103, *Survivability,* 10 June 1985.

FM 6-20, *Fire Support in the Airland Battle,* 17 May 1988.

FM 6-20-40, *Tactics, Techniques, and Procedures for Fire Support for Brigade Operations (Heavy),* 5 January 1990.

FM 6-20-50, *Tactics, Techniques, and Procedures for Fire Support for Brigade Operations (Light*), 5 January 1990.

FM 6-99.2, *U.S. Army Report and Message Formats*, 30 April 2007.

FM 7-0, *Training the Force,* 22 October 2002.

FM 7-15, *The Army Universal Task List,* 31 August 2003.

FM 10-67-1, *Concepts and Equipment of Petroleum Operations*, 2 April 1998.

FM 20-3, *Camouflage, Concealment, and Decoys,* 30 August 1999.

FM 34-2-1, *Tactics, Techniques, and Procedures for Reconnaissance and Surveillance and Intelligence Support to Counterreconnaissance,* 19 June 1991.

FM 34-60, *Counterintelligence*, 3 October 1995.

FM 55-30, *Army Motor Transport Units and Operations,* 27 June 1997.

FM 55-450-2, *Army Helicopter Internal Load Operations,* 5 June 1992.

FM 71-100, *Division Operations,* 28 August 1996.

FM 90-4, *Air Assault Operations,* 16 March 1987.

FM 100-8, *The Army in Multinational Operations,* 24 November 1997.

FM 100-13, *Battlefield Coordination Detachment (BCD),* 5 September 1996.

FM 100-15, *Corps Operations,* 29 October 1996.

GTA 05-08-001, *Survivability Positions,* 1 August 1993.

JP 1-02, *Department of Defense Dictionary of Military and Associated Terms*, 12 April 2001.

JP 3-0, *Joint Operations,* 17 September 2006.

JP 3-01.4. *Joint Tactics, Techniques, and Procedures for Joint Suppression of Enemy Air Defenses.* 25 July 1995.

JP 3-09, *Joint Fire Support*, 13 November 2006.

JP 3-09.1, *Joint Tactics, Techniques, and Procedures for Laser Designation Operations,* 28 May 1999.

JP 3-30, *Command and Control for Joint Air Operations*, 5 June 2003.

JP 3-50.2. *Doctrine for Joint Combat Search and Rescue.* 26 January 1996.

JP 3-50.21. *Joint Tactics, Techniques and Procedures for Combat Search and Rescue.* 23 March 1998.

JP 3-52, *Joint Doctrine for Airspace Control in the Combat Zone*, 30 August 2004.

JP 3-60, *Joint Targeting*, 13 April 2007.

STP 1-15D13-SM-TG, *Soldier's Manual and Trainer's Guide for MOS 15D, Aircraft Powertrain Repairer, Skill Levels 1 and 3*, 27 October 2004.

STP 1-15M13-SM-TG, *Soldier's Manual and Trainer's Guide for MOS 15M, UH-1 Helicopter Repairer, Skill Levels 1, 2, and 3*, 26 October 2004.

STP 1-15T13-SM-TG, *Soldier's Manual and Trainer's Guide, MOS 15T, UH-60 Helicopter Repairer, Skill Levels 1, 2, and 3*, 6 January 2005.

STP 1-15U13-SM-TG, *Soldier's Manual and Trainer's Guide for MOS 15U, CH-47D Helicopter Repairer, Skill Levels 1, 2, and 3,* 29 October 2004.

STP 1-93C1-SM-TG, *Soldier's Manual and Trainer's Guide, MOS 93C, Air Traffic Control, Skill Level 1*, 1 April 2002.

STP 1-93C24-SM-TG, *Soldier's Manual and Trainer's Guide, MOS 93C, Air Traffic Control, Skill Level 2, 3, and 4,* 2 April 2002.

STP 1-93P1-SM-TG, *Soldier's Manual and Trainer's Guide for MOS 93P, Aviation Operations Specialist Skill Level 1*, 1 October 2002.

STP 1-93P24-SM-TG. *Soldier's Manual Skill Levels 2/3/4 and Trainer's Guide, MOS 93P, Aviation Operations Specialist,* 1 October 2002.

TC 1-218, *Aircrew Training Manual, Utility Airplane C-12,* 13 September 2005.

TC 1-219, *Aircrew Training Manual Guardrail Common Sensor Airplane RC-12*, 3 June 2002.

TC 1-228, *Aircrew Training Manual, OH-58A Kiowa Helicopter,* 13 June 2006.

TC 1-237, *Aircrew Training Manual Utility Helicopter H-60 Series,* 27 September 2005.

TC 1-238, *Aircrew Training Manual, Attack Helicopter, AH-64A*, 23 September 2005.

TC 1-240, *Aircrew Training Manual, Cargo Helicopter, CH-47D,* 12 September 2005.

TC 1-248, *Aircrew Training Manual, OH-58D, Kiowa Warrior,* 12 April 2005.

TC 1-251, *Aircrew Training Manual, Attack Helicopter AH-64D,* 14 September 2005.

TC 1-600, *Unmanned Aircraft System Commander's Guide and Aircrew Training Manual,* 23 August 2007.

TM 1-1500-204-23-1, *Aviation Unit Maintenance (AVUM) and Aviation Intermediate Maintenance (AVIM) Manual for General Aircraft Maintenance (General Maintenance and Practices) Volume 1,* 31 July 1992.

TM 55-1560-307-13&P, *Operator, Aviation Unit and Aviation Intermediate Maintenance Manual With Repair Parts and Special Tools List for Extended Range Fuel System Army Model CH-47 Helicopter Part Number 855DSCC-D-0007-2 (NSN 1560-01-221-7600),* 11 December 1990.

Unit Movement Officer Deployment Handbook Reference 97-1, U.S. Army Transportation School, Fort Eustis, VA.

USASC, *Commander and Staff Risk Management Booklet,* 6 January 1999.

USASC, *Small Unit Risk Management Booklet*, 6 January 1999.

Index

B

C

Made in the USA
Monee, IL
04 November 2026

67387215R00144